D0793522

# THE HANDBOOK OF
# PEST CONTROL

# The Handbook
## of
# PEST CONTROL

*by*

*JAMES G. HAMM*

2   S

*Frederick Fell Publishers, Inc.*

*New York*

For information address:
FREDERICK FELL PUBLISHERS, INC.
386 Park Avenue South
New York, New York 10016

Library of Congress Catalog Card Number: 81-68913
International Standard Book Number: 0-8119-0331-1

MANUFACTURED IN THE UNITED STATES OF AMERICA

1  2  3  4  5  6  7  8  9  0

Published simultaneously in Canada by Fitzhenry & Whiteside Limited, Toronto

# CONTENTS

# THE HANDBOOK OF
# PEST CONTROL

# DISCLAIMER

You *alone* are responsible for determining whether a given pesticide can be legally purchased and applied by yourself, or whether it is restricted to use by professional pest control operators. Neither the author nor the publisher of this book is responsible for making pesticide legal-usage determinations for users of this book.

Research and evaluations regarding various pesticides is occurring continuously by the Environmental Protection Agency, universities, and other organizations. Consequently, pesticide legal-usage data—involving percentages and strengths, specific pests against which they can be used, areas where they can be applied, etc.—are changing constantly. At any given time, a certain pesticide may be either under evaluation and its status undetermined; restricted; banned; or available for general usage. It is virtually impossible to stay totally abreast of this changing situation.

The information in this book is general in nature. It is meant to inform you in *general terms* about pesticides and the pests against which they are used. Because the legal-usage status of pesticides is subject to change at any time, ONLY the pesticide CONTAINER LABEL can be depended upon to inform you of the latest restrictions and legal-usage changes, if any, involving a particular pesticide. Thus, you alone are responsible for abiding by the law when purchasing and applying any pesticide.

Neither the author nor the publisher is responsible in any way for any accidental poisonings of humans, pets, or wildlife that may occur from recommendations made in this handbook. Safety rules regarding the usage of pesticides are given in this book, and it is the reader's responsibility to abide by these rules exactly when applying or using pesticides.

Finally, when applying a pesticide, the reader is cautioned to use the strength stated on the pesticide container label for the particular pest involved. It must be pointed out here that the pesticide strengths and percentages recommended in this handbook are approximations in some cases, and are based on the author's own personal experience and upon a review of pest control literature—much of which is already outdated due to constantly changing pesticide regulations.

# PREFACE

No household, business establishment, nor other human habitat is immune to biological pests; nor is any flower garden, greenhouse, vegetable garden, orchard, or lawn immune to these pests: insects, spiders, ticks, mites, rats, mice, birds, snakes, etc. Many of these pests—particularly insects and ticks—have existed on earth millions of years *longer* than has man himself and, consequently, have adapted well to continued survival. Despite constant and often ingenious effort on his part, man never has been able to eliminate pests. Indeed, his *only* success lies in controlling them to some degree.

Pests commonly take up residence with man—causing filth and damage to his buildings and personal belongings, destroying his flowers and vegetables, consuming and contaminating his food, injuring his trees, shrubs, and lawns . . . even attacking the human body itself. Thus, many pests actually present a direct, or indirect, threat to man's own health and physical well-being. Even when they do not produce actual physical damage or threaten man's health, pests nevertheless cause annoyance and embarrassment. No one wants to live with them whether they are harmful or not.

As a professional public health biologist/entomologist, I serve as a pest control advisor and consultant to the general public. From this experience, I have discovered among the general public today a widespread lack of knowledge about pests and how to control them. I have also found that many people desire to know more about household, garden, and yard pests—and, particularly, how they *themselves* can control these pests safely, effectively, and economically. Thus, this handbook was written to fulfill this particular need.

3

However, the publication of this handbook is *not* meant to rob private pest control operators (P.C.O.s) of their means of livelihood. Rather, this handbook is meant to fulfill a particular need: to provide practical, easy-to-understand, do-it-yourself instructions to those people who desire to handle some of their pest control problems themselves.

In many cases, one can do an equally good job of controlling a given pest, or pests, and at considerable savings compared to hiring a private pest control firm. No person, however, can expect to handle each and every pest control problem that arises. Some pests present technical difficulties that are beyond the means and ability of the ordinary person. In such cases, hiring a P.C.O. is, of course, recommended.

Neither is this handbook meant to serve as a textbook of pesticides, of entomology (study of insects), nor of zoology (study of animals). Rather, its purpose is to serve as a *reference book*, containing specific and practical information and instructions related to most marketed pesticides—and to control of the more common forms of biological pests encountered by the average person. Always, my purpose is to present *practical* information that can be *applied*, directly or indirectly, by the reader.

Finally, in a world that is literally inundated with the printed word—magazines, books, pamphlets, brochures, advertising flyers, newspapers, etc.—I have made a diligent effort to keep this handbook brief and to the point. Hopefully, every statement and/or fact given will serve to further the reader's practical knowledge of home pest control. . . . In a word, this text is meant to be *used*. It is meant to be a practical home reference handbook to which the reader can refer again and again.

# HOW TO USE
# THIS BOOK

NOTE: You are urged to study carefully this section before using this handbook—and especially before applying any pesticide.

This handbook consists of two basic sections of text (Part I and Part II) plus an appendix. Part I covers *pesticides*, both in general and in detail. It provides all the information that you need about buying, handling, and applying any of the great number of available pesticides currently being sold to consumers in most countries of the world. The *Master List of Pesticides* is alphabetically arranged so that you can quickly find and read about any given pesticide. Also, a *General Usage Pesticide List* is included. It gives the broad-spectrum, general-usage pesticides that most readers will find to be of greatest value for their everyday pest-control problems.

Part II is devoted entirely to the various biological pests that trouble mankind. I have deliberately chosen *not* to arrange these pests in the customary groups; rather, I have presented them in simple alphabetical order. Most pest-control writers do in fact present pests in groups, such as: household pests; yard pests; greenhouse pests; garden pests; etc. However, these groups tend to be somewhat arbitrary and incomplete in most cases. A great many pests, for example, refuse to be grouped in neat and convenient categories and are, in fact, found in several different groups. Almost invariably any group of pests is likely to leave out one or several troublesome pest species. Therefore, the pests are listed alphabetically. Generally, the biology, habits, and importance of each pest is discussed, followed by a detailed discussion of recommended control measures.

Finally, the appendix contains the various tables, charts, and pesticide listings related to selecting, mixing, and applying pesticides. For example, a *Toxicity Comparison Chart* is provided, enabling you to tell at a glance just how toxic a given pesticide is. The *Dilution Tables* will help guide you in mixing concentrated liquid and dust pesticides to the desired strength.

### General Warning

*All* pesticides are toxic (poisonous) to man at least to some extent. Moreover, some pesticides are more toxic than others, and some are deadly poisons. Thus, everyone who handles or applies pesticides *must* take the necessary precautions to prevent poisoning of the applicator and/or those around him or her. Careful reading and following of the safety rules given below will help prevent the accidental poisoning of family members and pets by pesticides applied in or near the home.

### Safety Rules for Using Pesticides

1. *Never* apply any pesticide without first reading the container label. All pesticides are different, and some require special handling for maximum safety and effectiveness. *Only* the pesticide container label contains this specialized information.
2. Mix only as much pesticide as is needed for one job. This eliminates the problem of storage, or disposal, of unused pesticide. Dilution tables are provided in the appendix to enable you to mix the approximate amount of pesticide needed for a given problem.
3. Do not allow children or pets near when mixing or applying pesticides.
4. Always wear gloves, preferably *rubber*, when handling and applying pesticides.
5. Mix pesticides in a well-ventilated area—preferably out of doors.
6. Do not eat or smoke when mixing or applying pesticides.
7. Avoid getting pesticides on your bare skin. If you should spill pesticide on your skin, wash it off *immediately* with soap and water.
8. Especially avoid getting pesticide in your eyes, mouth, or nose. Avoid breathing dust or vapors from pesticides.
9. Never apply a pesticide directly to cooking utensils, dishes, or other household items that are handled frequently. Always remove all dishes, pots, and pans from cabinets and cupboards *before* applying a pesticide there.

10. When a liquid pesticide is used, inside or outside, always allow it to dry *completely* before human or animal contact with it begins.

11. Use the following special precautions when storing pesticides:
    A. Always store pesticides in their *original* containers—NEVER in an unlabeled container.
    B. Keep these containers tightly closed.
    C. Store these containers in a *locked* cabinet or room.
    D. Safely dispose of empty pesticide containers:
       1. Wash and rinse out each container.
       2. Punch holes in the container so that it cannot be used again.
       3. Bury empty containers in at least two feet of earth—or take them to an *active* city landfill for disposal.

12. Finally, if possible, keep the specific *antidote* on hand for the particular pesticide that you are using and/or storing.

REMEMBER: Pesticides generally are *safe* when they are handled and applied correctly. Most poisonings and deaths resulting from pesticides occur because of simple *carelessness*—such as allowing children to find and open pesticide containers. Rarely do accidental poisonings occur from the use of household pesticides when the container label directions and the above safety rules are followed exactly.

Pesticides are, in fact, much like fire: they can be both good *and* bad, depending upon how they are used. Use them correctly and treat them with respect, and they will serve you well.

# KEY TERMS

It is suggested that you familiarize yourself with the following key words. All of them are of prime importance in buying, mixing, and applying pesticides.

*Arachnid*—A class of arthropods: Class Arachnida. This class includes ticks, mites, and spiders: all of which have eight legs compared to six legs for insects; two body parts compared to three for insects; and usually no antennae which insects possess.

*Arthropod* (*arthro* = jointed + *pod* = foot or leg).—This phylum is a tremendous one in the animal kingdom. In fact it is the largest phylum both in sheer numbers and in types of organisms. It includes *all* jointed-legged animals with exoskeletons (skeleton located on the outside; that is, the hardshell animals). Some examples are crustaceans, insects, and arachnids.

*Baits*—several pesticides are available as baits—usually in granular or paste form. Baits are actually *lures* designed to attract the insect or other pest to *eat* them and, thus, consume the pesticide which usually kills quickly. An exception to this quick-kill bait rule is the anti-coagulant rodent baits. Baits should be placed at floor level—safely out of reach of children and pets. Baits should *never* be used in areas where food is prepared.

*Concentrate*—refers to the *amount*, or percent, of pure pesticide chemical present in a container, solution, or mixture. Thus "concentration" refers to the relative strength of a pesticide. For example, a "100% concentrate" contains nothing except pesticide and thus is *very*

8

concentrated and strong. Likewise, a "20% concentrate" contains only 20% pesticide chemical plus 80% mixer of diluent. Thus, it is said to be *less concentrated.*

*Diluent*—this term is derived from the word *dilute*, and refers to any substance or substances (usually liquid) used to dilute or weaken a pesticide to the desired concentration or strength. A diluent also serves as a *carrier* for the pesticide and thus makes for easier and more consistent application. Water, mineral oil, and diesel fuel are examples of liquid diluents commonly used with various pesticides.

*Dusts*—refer to dry mixtures of pesticides with (usually) an inert (inactive) powder. Dusts may be bought as concentrates and further diluted with dry powder diluent-carrier to the desired concentration. More preferably, dusts should be bought ready-mixed for use because it is difficult to mix dust concentrates to the correct strength. Dusts can be used on almost any surface or material without harm, but the visible powdery dust may be objectionable in some cases. Moreover, dusts tend to give good, long residual action—so long as they remain dry. When they become damp or moist, however, dusts tend to cake up and become ineffective. Wettable powders, on the other hand, usually remain effective *when used as dusts* despite becoming moist or damp. Dusts are particularly good for inaccessible and hard-to-reach areas, such as cracks, crevices, underneath cabinets and appliances, and inside walls.

*Ectoparasites*—external parasites of animals, such as fleas, lice, mites, ticks, and flies.

*Emulsion*—has three (3) ingredients: 1) *water*; 2) *solvent* (in which the pesticide itself is dissolved); and 3) an *emulsifier.* This emulsifier aids the dispersion of the solvent-plus-pesticide into very fine droplets throughout the water. The water, not the solvent, is the carrier in this case. Emulsions are necessary for pesticides of an oily or organic nature because these are naturally insoluble in water alone. An example of a common emulsion is: soap, water, and grease or oil. Since grease and oil will not mix with nor dissolve in water alone, soap acts as an emulsifier to chemically break up the oil or grease into tiny droplets that are dispersed evenly in the water-and-soap solution.

*Granules*—refer to pesticides made in granular (grain) form. In this form, the pesticide chemical is either coated *on* or impregnated *in* the granular carrier material. Granular pesticides are applied simply by scattering them throughout the area to be treated.

*Metamorphosis* (*meta* = change + *morph* = shape or form).—This term refers to the change of form that most insects and some arachnids go through to become adults; that is, egg, larva, pupa, and adult.

*Overwinter*—refers to a condition somewhat like the hibernation of animals in which insects and arachnids spend the cold months of the year.

However, when temperature and humidity are favorable, these pests often become active temporarily even during the cold months. Insects and arachnids living inside heated buildings may not overwinter at all, but, rather, remain active year-round.

*Residue*—refers to the *dry* and *fixed* pesticide that is left in place *after* the liquid diluent or carrier evaporates. Pesticide residues last for various lengths of time and *continue* to exert their killing action when insects crawl across or touch this *residual* layer. Thus, *residual action* refers to the effectiveness of the pesticide *following* its application.

*Solvent*—refers to any liquid used to dissolve another (usually dry) substance. Upon being dissolved, this substance is said to be "in solution." For example, when common table salt or sugar is poured into a glass of water, it is quickly *dissolved* by the water which is the *solvent*. Many pesticides likewise are dissolved in water, oils, alcohol, or some other appropriate solvent before being applied.

*Systemics*—refer to pesticides that can be given internally to animals to control *external* pests (ectoparasites). Foliar systemics are pesticides that are applied to the soil and, thus, to the roots of various plants. Plant, or foliar, systemics are transported biologically from the roots up to the green plant body and leaves where they poison sap-sucking insects.

*Tracking Powders*—are powder-form pesticides that usually are effective when mice have adequate food supplies and will not accept baits. Tracking powders are applied so that mice pick it up on their bodies. It kills when the mice groom themselves and thus ingest the powder.

*Vector*—in this case, refers to a pest—usually an insect, tick, or mite—that can, and often does, transmit disease organisms from animal to man. Mosquitoes, for instance, are by far the most important biological vectors affecting mankind's health. Mosquitoes bite various birds and animals, pick up numerous disease viruses (usually) and, in turn, transmit these directly to man through bites.

*Wettable Powder*—refers to a pesticide that is coated with an inert (inactive) powder to which a wetting agent is added. The wetting agent thus makes the powder-pesticide suspendable in water. The powder-pesticide does NOT dissolve in the water but is merely suspended in it. Thus frequent agitation is required to prevent the powder-pesticide from settling out.

The *wettable powder* form of a pesticide gives the greatest possible *residue* for that particular pesticide. Because this dry residue often is visible, however, wettable powders are somewhat limited for indoor use. On the other hand, wettable powders are highly recommended for outdoor use—particularly for treating lawns, trees, shrubs, ornamental plants, etc.

# PART I

# PESTICIDES

The term *pest* is a broad one indeed. It covers, literally, any living organism—insect, arachnid, bird, rodent, animal, weed, or even germ—that is a nuisance to mankind. The term *pesticide* (to kill pests) is thus a broad term also, and refers to any chemical agent used to kill, control, or repel any of the vast number and kinds of living pests. For practical purposes, however, the great number of pesticides available today are broken down and classified according to the general groups of pests that they are designed to control. The following list includes the major broad categories of pesticides.

*Acaricides*—chemicals used in particular to kill, control, or repel mites and ticks.

*Avicides*—chemicals used to kill, control, or repel birds.

*Fungicides*—chemicals used to kill or control fungi.

*Herbicides*—chemicals used to kill or control unwanted plants, weeds, and grasses.

*Insecticides*—chemicals used to kill, control, or repel insects. Because insects represent the largest single group of pests, insecticides likewise represent the largest single group of pesticides.

*Nematocides*—chemicals used to kill, control, or repel nematodes (parasitic worms).

*Rodenticides*—chemicals used to kill, control, or repel rodents (rats, mice, gophers, etc.).

More than 1200 pesticides currently are registered with the U.S. Environmental Protection Agency (E.P.A.). About 335 (approximately one-fourth) of these pesticides are classified as *insecticides*. Pesticide chemistry is thus a major industry both in the United States and throughout the world. Indeed, pesticides are one of mankind's essential tools— no, weapons!—as the global population rises, bringing with it increased demand for food and higher living standards. Daily we find ourselves combating and competing with literally *billions* of insects, arachnids, fungi, viruses, bacteria, weeds, and animals.

At present, our only hope of controlling, or at least containing, these colossal hordes of pests is through the judicious use of pesticides. Indeed, mankind's greatest single source of food—green plants—soon would be consumed by insects and other pests, thus bringing on global famine and starvation—were it not for our arsenal of pesticides. Likewise, the use of pesticides throughout the world has saved literally millions of human lives from killer diseases spread by insects and other pests.

Unfortunately, our heavy dependence on pesticides over the years has created some problems of its own—namely, contamination of the environment and increasing biological resistance developed by pests against the poisons used to control them. Consequently, it became necessary that we stop and evaluate these problems, and then take steps to correct and control them. Thus, with concern about air and water pollution, came the creation of the U.S. Environmental Protection Agency (E.P.A.) and other similar regulatory agencies throughout the world.

Continuous environmental impact studies of pesticides made by these agencies has led to a ban against the usage of some specific pesticides and to restricted usage of others. However, new, often safer, and in some cases, more effective pesticides have been developed to replace the banned and restricted chemicals. And so we continue to depend upon and use pesticides, from the big-time farmer who sprays thousands of acres of crops for insect and weed pests, to the greenhouse owner, to the truckfarmer, down finally to the individual who finds it necessary to treat his or her home, garden, yard, or other limited areas for a wide variety of biological pests.

Nor should the average person be afraid or hesitate to use pesticides when they are needed in the course of daily living. All pesticides, of course, are necessarily toxic (or poisonous). They have to be in order to fulfill their purpose. Toxicity is simply the ability of a substance to disrupt or interfere with the living chemical and physiological processes occurring continuously inside the cells and organs of living organisms—whether animal, plant, or germ. All pesticides, however, do not exhibit the same degree of toxicity to insects, animals, or humans. Some are far more toxic than others, some are deadly poisons, and a few are relatively harmless to humans and animals. Thus, the *relative toxicity* of some pesticides is high compared to other less poisonous pesticides.

Generally speaking, *all* insecticides are toxic to *all* insects to *some* degree. However, some insecticides are much more toxic and, thus, much more effective against a given pest. Choosing the right pesticide for a given pest is rather like a doctor choosing the correct, or best, antibiotic for an infection. The doctor has maybe ten or twelve antibiotics from which to choose. All are good, sound, effective medicines. . . . But, almost invariably, *one* or more of these antibiotics will be best for the particular patient at hand. And so the doctor chooses accordingly. The same is true about treating for pests. Drugs and pesticides have much in common. Both are in fact chemicals, and both can be good or bad, depending on *how* they are used.

Despite their inherent toxic nature, however, pesticides in the United States actually cause *fewer* accidental poisonings, both fatal and non-fatal, than do drugs and pharmaceutical products, according to the Food and Drug Administration (F.D.A.). Statistics published by the F.D.A. show that only 4.8% of all accidental poisonings in the United States resulted from pesticides, compared to 44.3% for drugs and pharmaceuticals and 16.3% for cleaning and polishing agents.*

*Source: Bulletin, National Clearing House for Poison Control Centers. U.S. Food and Drug Administration, Bureau of Drugs, H.E.W., May–June, 1974.

# USING
# PESTICIDES

*Please read the following paragraphs carefully.*

Again you are strongly urged to read carefully the *container labels* of all pesticides before applying them. The information given in this book is general in nature, and applies to pesticides as a group. However, each individual pesticide is *different*, and some require special handling, mixing, and application procedures for safety and effectiveness. ONLY the container label provides this special information.

As a general rule, pesticides applied *inside* the house should be WATER-BASED liquids, aerosols, powders, dusts, or baits. The reader should realize that oil-based sprays, or sprays containing a solvent other than water—such as xylene, alcohol, kerosene, or diesel—may *damage* or *discolor* surfaces to which they are applied. This is especially true of carpets, walls, furniture, and fabric materials. Again, you should consult the pesticide container label for any warnings or restrictions regarding application of the particular pesticide that you are using, both indoors and outdoors.

Pesticides should be applied in a safe manner. Children and pets must not be allowed to come into contact with pesticides.

Finally, you *alone* are responsible for determining whether a given pesticide can legally be purchased and applied by yourself, or whether its application is restricted to professional pest control operators. Remember that the evaluation of pesticides by the E.P.A. is occurring continuously, and at any given time, a certain pesticide may be either under evaluation; restricted; banned; or available for general usage. Neither the author nor the publisher is responsible for making individual pesticide legal-usage determinations. Indeed, the legal usage status of pesticides is changing constantly. This handbook simply informs you of the recommended and effective pesticides in each case, and it is your own responsibility to abide by the law when purchasing and applying any pesticide.

However, this handbook intends to help you use pesticides legally, as well as safely, effectively, and economically. Unlike most pest control books, this text lists a large selection of control pesticides for almost every pest species. This was done for a purpose. Because the legal status of many pesticides is changing almost continuously, the large selection provided here in each case makes it possible for you to always find at least *one* that is: (1) legal to use in your particular case and (2) available for purchasing.

In some instances this handbook lists recommended pesticides by a trade name or name-brand, indicated by capitalization or the symbol®, and in other cases the generic chemical name is given. For example, the generic chemical, chlordimeform, is listed in several instances; in other cases, two trade names for it are listed: Galecron and Fundal. Another example is the generic chemical warfarin which, among other trade names, is sold as D-Con. When selecting a pesticide from your recommended list, remember that the generic chemical itself is the active agent in every case, not the trade name. Thus, when chlordimeform, for example, is recommended for ticks, you may select *any* name-brand product that contains chlordimeform, including the two specific products mentioned—Galecron and Fundal. The same is true of any name-brand product containing warfarin, as well as all other generic chemicals. Consequently, before you buy a product, always check the container label to be sure it contains the generic chemical (active agent) that you seek.

# SELECTING
# A PESTICIDE

When faced with a pest control problem that you wish to handle yourself, it is first necessary to assess the problem. Exactly what pest, or pests, are present? How bad is the infestation? Is it new or long-standing? Is it widespread or merely limited to a small area? Can you adequately handle this control problem yourself?

Don't be afraid to ask yourself an honest question . . . then listen to your inner self for an answer. You will *know* whether you are really capable of handling the problem or not.

Assuming that you decide to handle the problem yourself, you now must decide which method, or methods, to use. For example, should you actually kill the pests—or merely repel them? At this point, you should realize that killing the pest species involved in a problem is not always the best solution to the problem. In some cases, depending upon the type of pest involved, killing is definitely not indicated nor recommended—and may, in fact, be illegal. Rather, repelling the pests may be the best solution. When repelling rather than killing the pest is recommended, the Control Section for that particular pest species will discuss the methods.

Another consideration is the pesticide, or pesticides, to use for the particular problem at hand. After checking the handbook for the recommended control method and/or pesticide(s), you must determine whether these poisons are available to you. If they are not available to you, you will be forced to consider alternative pesticides which may, or may not, be quite as effective. Also, the *toxicity* of the pesticide recommended should be considered.

Is it highly toxic? Are there other almost equally good, but less toxic, pesticides that you could use? Would applying a highly toxic pesticide be likely to endanger members of your family, pets, wildlife, or even your neighbors? Finally, can one of the general-use/broad-spectrum pesticides be used effectively? If so, you will find it advantageous to purchase one of these general-use pesticides because it can be applied to control a number of different pests.

Generally speaking, when you select a pesticide, look for: (1) low to moderate toxicity; (2) general availability; (3) wide-spectrum usage; and (4) specificity against the particular pest that you wish to control.

It is suggested that, after assessing the problem as discussed above, you now turn to the pest species in Part II of this handbook. Read all the information given, including all the recommended control measures for that pest. If pesticides are to be used:

1. Make a list of *all* the recommended pesticides for that pest.
2. Take this list with you to buy the pesticide.
3. If possible, go first to a farm supply or garden supply store, since these establishments usually stock a wide variety of pesticides.
4. If your first-choice pesticide is not available, simply drop down to the next one on your list—and so on until you find one that is in stock.
5. If neither farm nor garden supply stores are convenient, try supermarkets, large discount department stores, and drug stores.
6. If still unable to locate a suitable pesticide, turn to the yellow pages of your phone book. Look under PESTICIDES. If, in this case, the supplier only sells in large or bulk quantities, you may wish to ask a friend or neighbor to share the cost of purchasing a larger-than-needed supply, which nevertheless can be used by the both of you over a period of time.

# GENERIC
# VERSUS NAME-BRAND
# PESTICIDES

The reader should understand that, with few exceptions, the pesticides discussed in this handbook are the *pure* forms of the chemicals. That is, these are the *generic* forms and are not to be compared with the name-brand products commonly advertised and sold in supermarkets. These commercial products usually contain a mixture of several ingredients that generally are less concentrated and, thus, less potent than the pure or generic chemicals which you can formulate to the desired strength.

But whether you select a generic chemical itself or a name-brand mixture *containing* the generic chemical, it is important in any case to remember that the generic chemical itself is the active pesticidal agent— *not* the product name. There are, of course, only a certain number of generic chemical pesticides available. On the other hand, there is an almost unlimited number of name-brand products available. Thus, the important thing to check in selecting a name-brand product is whether it *contains* the actual *generic chemical* that is recommended for the particular pest that you wish to control.

To determine this you must read the container label—and look closely at the list of ingredients. For example, if you wish to use warfarin to kill rats and mice, you can select any of a number of name-brand rodenticides whose active ingredient is warfarin. For instance, Raze Rat and Mouse Bait and D-Con are only two of the many name-brand products that contain warfarin.

On the other hand, you may prefer to buy the generic chemical pesticide itself in concentrated form and dilute it to the recommended strength.

18

# TOXICITY OF PESTICIDES (LD$_{50}$ VALUES)

How do scientists measure and compare the toxicity of the various pesticides? That is, how do they compare the toxicity of a particular pesticide to others? Scientists use a standard laboratory measuring procedure, called the "LD$_{50}$ method", to compare the toxicity of pesticides. LD$_{50}$ means, very simply, *Lethal Dose—50%*. This refers to the exact amount of a given pesticide, measured in milligrams (mg) per kilogram (kg) of body weight that is required to kill 50% of the test animals or insects on which it is tested. Thus, science uses a standard and repeatable method to measure the toxicity of pesticides. This method provides for the immediate and direct comparison of toxicity between two or more pesticides.

At this point the reader should realize that a milligram is a very small amount indeed. It is one twenty-eight thousandth of an ounce. That is, one ounce contains approximately 28,000 milligrams. A kilogram, however, is more than twice the weight of a pound. One kilogram (kg) equals about 2.2 pounds.

Now, let's compare the relative toxicity of two different pesticides—parathion and diazinon. The LD$_{50}$ value of parathion, an extremely toxic chemical, is 3.6. That is, 3.6 milligrams of parathion per kilogram of body weight killed at least 50% of the animals (usually rats) on which it was tested. This compares to a whopping LD$_{50}$ value of 285 for diazinon, a moderately toxic pesticide. By comparison, the LD$_{50}$ value of malathion, a very *low* toxicity pesticide, is 1000.

Since, obviously, pesticides are tested on animals and not on humans, how does this animal toxicity, or LD$_{50}$ value, translate to humans?

What do these values *mean* in terms of human toxicity? Generally speaking, only a *few drops* of any pesticide with an LD$_{50}$ value of *less* than 5 (parathion, remember, is 3.6) most likely will kill a 150 pound human if taken orally (internally).

Now let's compare this high toxicity level of parathion to the moderately toxic pesticide, diazinon, and to the very low toxic malathion—all in human terms. With an LD$_{50}$ value of 285, the approximate amount of diazinon required to kill a 150 pound human is between *one* and *two* *tablespoonfuls*—compared to only a few drops of parathion. For a child or a person weighing substantially less than 150 pounds, however, even less diazinon or parathion would be likely to prove fatal—since the toxicity level depends directly upon body weight. As for malathion with an LD$_{50}$ value of 1000, a very large quantity—approximately *one pint* taken orally—is required to kill a 150-pound human.

Most readers will want to compare the toxicity of the various pesticides available today. The table below makes it possible for you to determine at a glance just *how* toxic a given pesticide is to human beings. It is suggested that you familiarize yourself with this table. It has been adapted from Hayes (1963)* and is a good general guide for estimating the probable lethal or fatal dose of a pesticide, based on its tested LD$_{50}$ value, for an adult human being weighing approximately 150 pounds. NOTE: This table relates to *human* toxicity—not necessarily to pest toxicity.

REMEMBER: The *lower* the LD$_{50}$ value of a pesticide the *higher* is its toxicity.

| Pesticide Toxicity Rating | Acute Oral LD$_{50}$ for any animal (*mg/kg*) | Probable Lethal Dose (*Oral for a 150 lb. Human*) |
|---|---|---|
| Extremely Toxic | Less than 5 | A few drops |
| Highly Toxic | 5 to 50 | A "pinch"; or 1 teaspoonful |
| Moderately Toxic | 50 to 500 | 1 teaspoonful to 2 tablespoonfuls |
| Low Toxic | 500 to 5,000 | 1 ounce to 1 pint; or 1 lb. |
| Non-Toxic | 5,000 to 15,000 | 1 pint to 1 quart; or 2 lbs. |

NOTE: The appendix in back contains toxicity comparison tables for most commonly used pesticides. These tables allow you to quickly determine the relative toxicity (high, moderate, low, etc.) of a particular pesticide.

*Hayes, W. J., Jr. Clinical Handbook on Economic Poisons. U.S. Public Health Service.

# PHYSIOLOGICAL ACTION OF PESTICIDES

Pesticides exert their killing action by biochemically and physiologically altering the living processes going on inside the target pest's body—whether it be insect, spider, bird, rat, or whatever. Thus it follows that, to exert its killing action, the pesticide first must get *inside* the pest's body. Usually this is a gradual, even slow, process except in those cases where the pests actually eat the pesticide or breathe it into their bodies. Unfortunately, many people, after applying an insecticide or pesticide, are disappointed and perturbed when they see the pests which they were trying to kill still moving about several hours, or even days, later. However, assuming that the appropriate pesticide was applied—correctly—and at adequate strength, this *delayed* killing action is, in most cases, normal and to be expected. Such results do *not* necessarily mean that your pesticide is not working—but, rather, that the slow chemical changes that cause death inside the pest's body have not yet had time to occur.

On the other hand, a few pesticides do in fact produce quick knock-downs and quick kills. But this is more the exception than the rule. Thus, in many cases, it is necessary to *remain patient* while the applied pesticide does its silent work.

Most pesticides are NERVE poisons. Chemically, they interfere with and disrupt the normal functioning of the nervous system of insects and animals. Since muscles control body movement and breathing—and since muscles, in turn, are controlled by nerves, it follows that disruption of the nervous system also disrupts the muscular system. This, in turn, disrupts body movement and breathing, leading usually to paralysis and suffo-cation from lack of oxygen taken in by the paralyzed breathing muscles. Mild or moderate disruption of the nervous system produces only symp-toms of poisoning in both insects and humans. Severe disruption produces convulsions, paralysis, and finally, death.

# PROTECTING
# HONEYBEES
# FROM INSECTICIDES

Honeybees serve a very useful and beneficial agricultural purpose: the cross-pollination of various field crops, flowers, and plants. Thus, honeybees generally are considered valuable and helpful insects rather than pests. In most cases, honeybees are protected from insecticides. The suggestions given below will help to protect honeybees from insecticides.

1. Apply the chemical in the safest possible manner. Use ground equipment rather than aerial or high-volume foggers.
2. Avoid using dust-form pesticides in great volume.
3. Make insecticide applications early in the morning or late in the day to avoid killing bees in flight.
4. Avoid treating crops, flowers, or other plants that are in *bloom*.

The tables that follow will prove helpful in selecting insecticides that are less toxic to honeybees. These lists and suggestions have been adapted from the University of Minnesota Agricultural Extension Service, Bulletin #387.

### Insecticides Highly Toxic to Honeybees

acephate (Orthene)
arsenic compounds
azinphosmethyl (Guthion)
carbaryl (Sevin)
carbofuran (Furadan)
chlorpyrifos (Dursban; Lorsban)

diazinon (Spectracide)
dichlorvos (DDVP; Vapona)
dimethoate (Cygon)
EPN
fenthion (Baytex)
heptachlor

lindane
malathion
methamidophos (Monitor)
methidathion
methomyl (Lannate)
methyl parathion
mevinphos (Phosdrin)

monocrotophos (Azodrin)
naled (Dibrom)
parathion
phosmet (Imidan)
phosphamidon (Dimecron)
stirofos (Gardona; Rabon)

## Insecticides Moderately Toxic to Honeybees

Abate
chlordane
crotoxyphos (Ciodrin)
coumaphos (Co-Ral)
DDT
demeton (Systox)

disulfoton (Di-Syston)
endosulfan (Thiodan)
Endrin
oxydemetonmethyl
phorate (Thimet)
ronnel (Korlan)

## Insecticides With Low Toxicity to Honeybees

allethrin
Aramite
bacillus thuringiensis
binapacryl (Morocide)
methoxychlor

Morestan
nicotine
Omite
ovex (Ovotran)

# ANTIDOTES
# TO PESTICIDES

An antidote is a drug or other substance which counteracts a poison in a human or animal. Pesticides, being poisons, have effective antidotes in most, but not all, cases. Unfortunately the recommended antidotes to most pesticides are not readily available to the general public; rather, they must be administered by a physician. Since pesticides are, in most cases, nerve poisons, they require special drugs as antidotes.

What should you do in case someone is accidentally poisoned by a pesticide? You are urged to read carefully the following suggestions which relate to medical aid for anyone who is accidentally poisoned by a pesticide:

1. If the pesticide container is available, *read the label* for emergency first aid measures.
2. Keep *visible* near your phone the number of the nearest POISON CONTROL CENTER. These centers have been created specifically to handle poisoning emergencies. Poison control centers handle *all types* of poisonings—not just pesticide poisonings.
3. Determine which *chemical group* the pesticide belongs in: botanicals, organophosphates, minerals, chlorinated hydrocarbons, carbamates, or fumigants. Most pesticides are listed in this handbook under *one* of the above chemical groups. This will be very important to the poison control center in determining which antidotes to use or to otherwise help the victim.
4. Tell the poison control center the exact *name* of the pesticide, the *active ingredients* (from the label), and the *chemical group* to which it belongs.

5. Do *exactly* what the poison center personnel tell you to do. Then take the victim to the nearest doctor at once.
6. Tell the doctor the pesticide name, the active ingredients, and the chemical group to which it belongs.
7. If the victim begins to *turn blue* and/or *ceases to breathe*, ADMINISTER ARTIFICIAL RESPIRATION *at once* (mouth-to-mouth or by pressing rhythmically on the victim's back above the lungs). Continue this artificial respiration until breathing is restored.

A person who has been poisoned by a pesticide usually will show symptoms within a few minutes to as much as a *few hours*. As a rule, the milder the symptoms and the slower their appearance, the less severe is the poisoning. In any case, death from pesticide poisoning usually results directly from *respiratory failure* due to paralysis of the breathing muscles. Thus, should the victim begin to turn blue or cease to breathe, it is IMPERATIVE that you administer artificial respiration *continuously* until breathing is restored or until medical help is at hand.

# SYMPTOMS OF PESTICIDE POISONING

Sometimes, particularly with small children or pets, you may not actually know that accidental poisoning has occurred. Thus, it is wise to familiarize yourself with the *general symptoms* of pesticide poisoning. Symptoms for each of the three most widely used groups of pesticides are summarized as follows:

*Organophosphates*

*Mild*—blurred vision; headache; dizziness; weakness; and anxiety. (A victim may show one or more of these symptoms.)

*Moderate*—uncontrolled watering of the eyes; salivation; nausea and/or vomiting; sweating; muscular tremors; abdominal cramps. (One or several of these symptoms may be present.)

*Severe*—difficulty in breathing; diarrhea; cyanosis (blue-colored skin); pinpoint eye pupils; loss of kidney and/or bowel control; convulsions; and coma.

*Carbamates*—pinpoint eye pupils; tightness in the chest; pain above the stomach; salivation, nausea and/or vomiting; diarrhea; excessive sweating; weakness and tiredness; muscle incoordination. (One or several of these symptoms may be present.)

*Chlorinated Hydrocarbons*—respiratory difficulty; cyanosis (blue-colored skin); nausea and/or vomiting; nervousness and restlessness; tremors; apprehension (fear and anxiety); coma. (Again, one or more of these symptoms may be present.)

REMEMBER: Pesticides *are safe*—if used as directed and the general safety rules given in this handbook are followed. Accidental pesticide poisoning almost always results from simple carelessness—or failure to follow the specified safety rules.

# E.P.A.
# PESTICIDE
# RESTRICTIONS

Residual and long-term cumulative effects of pesticides in the environment, along with growing concern about air and water pollution, led finally to creation of the U.S. Environmental Protection Agency (E.P.A.). *All* marketed pesticides now must be registered with the E.P.A. by their manufacturers.

Since October, 1977, all pesticides have been classified by E.P.A. into two broad categories: (1) RESTRICTED USAGE and (2) GENERAL USAGE. "Restricted Usage" refers to those specific pesticides which can be applied legally *only* by certified pest control operators or by someone working directly under the supervision of a P.C.O. "General Usage" pesticides, however, may be applied by anyone at any time without restriction.

# MIXING
# LIQUID
# PESTICIDES

Most pesticides are concentrated. Thus, they will need to be diluted to proper strength before using. This is true whether the pesticide is a dry form or a liquid which must be mixed with a liquid solvent or carrier. Some of these pesticides are soluble in water, which can be used as the diluent or carrier. Others, however, will not dissolve in nor mix with water.

Let's say that you have a pest control problem that calls for a pesticide which will *not* dissolve in water. What to do? First, check the supply shelf where you purchase pesticides to see whether there is a dust, wettable powder, bait, or granule form of this recommended pesticide available. Also check the handbook to see whether there is another equally good pesticide that you can substitute. In either case, a solution to your problem is provided.

If, for some reason, however, you must use the liquid form that is not soluble in water, it is necessary that you obtain a suitable solvent for this pesticide. As a rule, all pesticides that do not dissolve in water *will* dissolve in any of the following *organic* solvents—all of which are commonly used to dilute pesticides.

| | |
|---|---|
| acetone | fuel oil # 1 |
| benzene | diesel fuel |
| cyclohexanone | kerosene |
| ortho-dichlorobenzene | mineral oil |
| xylene | |

It is recommended that, if at all possible, you use one of the last four solvents. These are: fuel oil #1, diesel fuel, kerosene, or mineral oil.

These four solvents are all equally good. Moreover, they generally are safe, cheap, and readily available. However, in the unlikely event none of these four solvents can be obtained, then you should of course resort to any of the other solvents listed above. See the appendix for dilution tables.

# WHEN TO RE-TREAT
# FOR PESTS

Some pest control problems may require that you re-treat one or more times following the first application of pesticide. This is particularly true of many insects, ticks, and mites—all of which go through a development process called *metamorphosis* (change of form). That is, any time that you treat for these particular pests, very likely there will be present all stages of the pest's life cycle.

If the pest is the type that undergoes *complete* metamorphosis with four different life-cycle stages (egg, larvae, pupae, and adult), the pesticide usually will kill *only* the adults and, perhaps, a few larvae and pupae. Chances are the eggs that are present and incubating will be affected very little, if at all. Thus, in a few days, a new brood of pests will have hatched—and, thus, it may *appear* that your pesticide is not working. However, if the pesticide is the *residual* type, it will *continue* to work and kill the newly-hatched pests as they move across it over a period of time. Therefore, you should be patient and wait for results as was suggested earlier.

When then *do* you re-treat?

If, after ten to fourteen days, you still see *adult* pests in fairly large numbers, then perhaps you should re-treat with pesticide. If this application does not prove satisfactory after a few days, then you should consider a third application. Finally, if after the *third* application of pesticide you *still* see adult pests in bothersome numbers, then you should consider switching to a different pesticide. Start all over again and continue treating at one to two week intervals until the pests are eradicated or otherwise under control.

# BIOLOGICAL
# RESISTANCE
# IN PESTS

What about biological resistance to pesticides in pests, particularly in insects? What does this mean? It means simply that a given pest—say the German cockroach—can and sometimes does develop resistance to the poisons used to control it. This is especially true when *one* particular poison is used over and over again in the same area for long periods of time. Resistance is usually noticed by *decreased* effect of the pesticide; that is, the same amount of poison no longer kills as many insects. We should understand here, however, that resistance is *not* the same thing as immunity.

*Immunity* refers to the body's ability to develop antibodies to germs, viruses, and other disease organisms. Resistance, however, involves fundamental and basic changes in the insect's *genes*—usually the result of a *mutation* (change) brought about by continued usage of a given pesticide. That is, continuous usage of a given pesticide in the same area for long periods of time seems to bring about these changes in insect genes. The mutated genes then instruct the insect body cells to build new *enzymes* (complex chemicals inside all living cells) that are able to attack, chemically bind up, and de-activate the pesticide, thus rendering it harmless.

At first, only a few insects in a given population undergo genetic mutation from pesticide pressure. But these insects produce offspring with the same genetic changes present. The offspring, in turn, produce more progeny with still the same resistant genetic changes present. And so on, until all the insects present in the population are resistant to the given poison being used to control them.

What happens when resistance develops in an insect, or other pest, population? Fortunately, pests only develop resistance to *one* insecticide at a time. This, of course, leaves an arsenal of other pesticides which then can be applied against the resistant pests.

Nor does resistance against a certain pesticide develop world-wide all at once—unless, of course, it is used world-wide at the same rate for long periods of time. Rather, resistance tends to develop *locally*. For example, a mosquito species such as *Culex pipiens* which may have developed resistance to, say, malathion in California, will *not* be resistant to malathion in, say, Massachusetts—unless, of course, this chemical has been applied there continuously for a long period of time.

For the average person concerned with limited pest control problems on a small scale, resistance is not a problem in most cases. Usually you will eliminate the pest problem long before resistance could develop. The only problem, of course, is that the pest you are attempting to eradicate with a given pesticide may have developed resistance elsewhere. This is, in fact, somewhat likely with cockroaches, for example. Because cockroaches are so widespread and persistent, and because they are continually bombarded with a whole arsenal of poisons, they can and do become resistant to one or more of these pesticides from time to time. And since, moreover, cockroaches are highly mobile creatures—often traveling across country in food and furniture and other ways—a resistant strain *could* indeed end up in your kitchen.

What to do? Simply look for the effectiveness of the particular pesticide that you choose to use. If, after adequate time, this pesticide seems ineffective, then simply select another one from the list of recommended pesticides. There is more than one way to kill a cockroach—or any other pest!

# FUNCTIONAL PESTICIDE GROUPS

As discussed earlier, pesticides are classified according to the *types* of pests they are intended to control, such as insects, rodents, birds, weeds, etc. The lists provided below will help you to determine at a glance which pest group a given pesticide belongs in; that is, which pest (or pests) it is meant to control.

In some cases, a given pesticide may be listed in more than one group. This means, of course, that it can be used against more than one pest.

### Avicides

This list includes only the most widely used and listed chemicals against bird pests. Some of these are repellents rather than outright poisons. In fact, you will find it advisable in most bird pest cases to use an effective repellent rather than a poison because many bird species are protected by law in most states. Moreover, by actually killing birds, you may incur the wrath of various bird-loving groups.

| | |
|---|---|
| Avitrol | methiocarb (Mesural) |
| Baytex | Ornitrol |
| Bird Tanglefoot | Rid-A-Bird |
| endrin | Roost No More |
| fenthion (Baytex; Entex) | strychnine |
| For The Birds | |

## Acaracides/Miticides

Ticks and mites are arachnids, not insects. But, like many insect pests, ticks and mites are very common, persistent, destructive, and even dangerous. They are often difficult to control, and usually require special pesticides known as *acaracides* (for ticks) and *miticides* (for mites). The table below lists the more commonly used acaracides and miticides.

| | |
|---|---|
| Aramite | isothioate |
| chlorbenside (Mitox) | lindane |
| chlorobenzilate (Acarben) | malathion |
| chlordimeform (Galecron; Fundal) | Mitox |
| cyhexatin (Plictran) | Meta-Systox-R |
| Cygon | Morestan |
| diazinon (Spectracide) | Morocide |
| dicofol (Kelthane) | Omite |
| dinocap (Karathane) | ovex (Ovotran) |
| Dimecron | Pentac |
| dioxathion (Delnav) | phosmet |
| Dursban | Prolate |
| endosulfan (Thiodan) | Sevin (carbaryl) |
| EPN (phenyl phosphorothioate) | tetradifon (Tedion) |
| ethion | toxaphene |
| fenson | Trithion |
| Genite | Vendex |
| Imidan | Zectran |

## Anti-Coagulants

These are a special group of rodenticides that kill by interfering with the normal blood-clotting mechanism of animals, thus causing slow and continuous internal hemorrhaging. Anti-coagulants tend to be slow-acting and require a number of feedings to become effective. For this reason, they present very little danger to humans.

| | |
|---|---|
| coumachlor | Fumasol-G |
| coumatetralyl (Racumin) | pindone (Pival) |
| chlorophacinone (RoZol and other brand-name products) | Pivalyn |
| | PMP |
| diphacinone (Promar and other brand-name products) | warfarin (D-Con, Prolin, and other name-brand products) |
| Fumarin | |

### Defoliants

Defoliants are chemicals that hasten leaf fall from plants and trees. They are especially useful in agriculture for removing the leaves of various crop plants, thus facilitating harvest of the crops. Defoliants are also used in war zones to remove dense foliage from trees and plants, thus exposing enemy troops and equipment.

DEF®
disodium octaborate tetrahydrate
Folex

magnesium chlorate
paraquat
sodium chlorate

### Fumigants

These are *gases* that are used to fumigate buildings, houses, underground burrows, subfloors, between walls, and other hard-to-reach places. Usually fumigation as a pest-control method is employed only in very difficult or special cases. Fumigation is especially dangerous to mankind and animals. In most cases, it should be left to professional pest control personnel. The more common members of the fumigant group of pesticides are listed below:

acrylonitrile
carbon tetrachloride
carbon disulfide
chloropicrin (tear gas)
ethylene dibromide
ethylene dichloride
ethylene oxide

hydrogen cyanide
methyl bromide
naphthalene
PDB (paradichlorobenzene)
phosphine (Phostoxin)
sulfuryl fluoride (Vikane)

### Fungicides

Strictly speaking, fungicides are chemicals designed to kill fungi. They also may be used to control some types of bacteria. Fungicides are especially useful for treating plants and crops for bacterial and fungal conditions such as: root rots, smut, gall, rusts, wilts, leaf blights, and storage rots.

anilazine
basic copper sulfate
binapacryl (Morocide)
Captafol (difolatan)
Captan
carboxin
ceresan
chloroneb
chlorothalanil

chromium compounds
copper dihydrazine sulfate
copper oxychloride
copper oxychloride sulfate
copper zinc chromate
cupric (copper) carbonate
cuprous (copper) oxide
dinitrocresol
dinocap (Karathane)

Ferbam
Folpet
HCB (hexachlorobenzene)
maneb
mercury compounds
Nabam
oxycarboxin
PCNB (pentachloronitrobenzene)
PCP (pentachlorophenol)

sodium compounds
sulfur
terrazole
Thiram
Vapam
zinc
Zineb
Ziram

## Herbicides

Herbicides are chemicals used to kill unwanted vegetation, grasses, and weeds in crops, on ditchbanks, rights of way, in gardens, along fence rows, in recreational areas, yards and turfs, and other areas. *Selective* pesticides kill only the harmful weeds and grasses. *Non-selective* herbicides kill *all* vegetation present.

acrolein
alachlor
allidochlor (CDAA; selective)
amitrole (selective)
ammonium nitrate (non-selective)
ammonium sulfate (non-selective)
ammonium thiocyanate (non-selective)
ammonium trioxide (non-selective)
arsenic trioxide (non-selective)
arsinic acid
arsonic acid
atrazine (selective)
barban (selective)
benefin
bensulide
bromacil (selective)
cacadylic acid
CDEC (selective)
chloramben
chlorpropham (selective)
copper sulfate (selective)
cypromid
dalaplon (selective)
DCPA
diallate (selective)
dicamba
dichlobenil (selective)
dinoseb (non-selective)
diuron (selective)

diquat (non-selective)
diphenamid (selective)
DNOC (non-selective)
DSMA
endothall (selective)
EPTC (selective)
fenac
iron sulfate
linuron (selective)
MCPA (selective)
monuron (selective)
MSMA
NPA (naptalam; selective)
paraquat (non-selective)
PCP (pentachlorophenol; non-selective)
pebulate (selective)
petroleum oils (diesel, fuel oil, kerosene, motor oil; non-selective)
picloram (selective)
propanil (selective)
propazine (selective)
propham (selective)
prometryne (selective)
pyrazon
pyridine (selective)
silvex (selective)
simazine (selective)
sodium arsenite

sodium borate (non-selective)
sodium chlorate (non-selective)
sodium metaborate (non-selective)
sodium tetraborate (non-selective)
TCA (selective)
terbacil (selective)

terbutol (selective)
trifluralin
two, four, five-T (2,4,5-T; selective)
two, four-DB (2,4-DB; selective)
two, three, six-TBA (2,4,6-TBA; selective)

**Insecticides**

Insecticides constitute the single largest category of pesticides. Insecticides are further classified by the *chemical group* to which they belong. That is, a given chemical group contains insecticides all of which have a similar chemical structure and, in most cases, similar actions against pests. Moreover, the members of each chemical group of insecticides usually exert basically the same effects in the environment. The primary chemical groups of insecticides are discussed below.

***Botanicals*** (plant-derived insecticides)—Pyrethrins are the best example of natural or plant-derived insecticides. Generally, pyrethrins knock down insects quickly, but they do not always kill. For this reason, they usually are mixed with one or more other insecticides before being used. Pyrethrin toxicity to humans is very low, making them among the safest of all insecticides. *Synthetic* (man-made) pyrethrins are similar to natural pyrethrins, but generally show greater killing power than the natural products. The most important members of the botanical group of pesticides are listed below:

allethrin (synthetic)
barthrin (synthetic)
dimethrin (synthetic)
Neopynamin (synthetic)
nicotine

pyrethrum (pyrethrins)
resmethrin (synthetic)
rotenone
ryania
sabadilla

***Carbamates***—This group is characterized by carbamic acid ($CO_2NH_3$). This is a relatively new group of insecticides that may eventually replace the organophosphates. Sevin and Baygon, in particular, are two commonly-used and effective general-purpose insecticides of this group. The most common members of the carbamate group of pesticides are listed below:

aldicarb
BUX® (phenyl methylcarbamate)
carbaryl (Sevin)
carbofuran
dimetilan

formetanate
methiocarb
methomyl
propoxur (Baygon)
Zectran (methylcarbamate)

***Chlorinated Hydrocarbons***—This group of insecticides are characterized by the chemical elements hydrogen, oxygen, and chlorine. These pesticides were used widely from the early 1940's through the 1960's. Chemically they are very stable (do not break down or degrade) and, thus, have accumulated gradually in the environment—particularly in soil, animal and fish tissues, plants, and water. Continued wide-spread usage of these insecticides has been restricted by the E.P.A. The more common members of the now generally restricted chlorinated hydrocarbon group of pesticides are listed below:

| | |
|---|---|
| Acaralate | endosulfan |
| Acarol | endrin |
| aldrin | heptachlor |
| BHC (benzene hexachloride) | Kepone |
| chlordane | lindane |
| chlorobenzilate | methoxychlor |
| DDD (see TDE) | mirex |
| DDT (dichloro-diphenyl-trichloro- | Perthane |
| ethane) | TDE (DDD) |
| dicofol | toxaphene |
| dieldrin | |

*Minerals*—Minerals were among the very first pesticides known to and used by man. Generally, however, they do not show very good or rapid insect killing action. Basically minerals as insecticides have been replaced by newer and more effective chemicals. The most important members of the mineral group of pesticides are listed below:

| | |
|---|---|
| arsenic | phosphorus |
| borax | sodium |
| copper | sulphur |
| dry diluents | zinc |
| lead | |
| petroleum oils (kerosene, diesel, fuel oil, etc.) | |

*Organophosphates*—This large, very useful, and effective group of chemicals generally has replaced the chlorinated hydrocarbons for wide-scale usage. In the environment, this group of insecticides tends to degrade or break down rapidly into harmless components. This group also tends to break down readily in the presence of water. Thus, it presents much less environmental danger than does the chlorinated hydrocarbons. The most important members of the organophosphate group of pesticides are listed below:

| | |
|---|---|
| Abate | bromophos |
| azinphosethyl | bromophosethyl |
| azinphosmethyl | carbophenothion |
| Bidrin | chlorfenvinphos |

chlorpyrifos (Dursban)
Ciodrin
coumaphos (Co-Ral)
demeton
diazinon (Spectracide)
dicaphthon
dichlorvos (DDVP; Vapona)
dimethoate (Cygon)
disulfoton (Di-Syston)
EPN (phenyl phosphorothioate)
ethion
famphur
fenthion (phosphorothioate); Baytex; Entex
formothion
Imidan
malathion

menazon
methidathion
methyl parathion/ethyl parathion
mevinphos
naled (Dibrom)
Nemafos
oxydemetonmethyl
phorate
phosalone
Phosdrin
Rogor
ronnel (Korlan)
Ruelene (methylphosphoramidate)
TEPP (tetraethyl pyrophosphate)
tetrachlorvinphos (Gardona)
trichlorfon (Dipterex)
Zytron (phosphoramidothioate)

### Nematicides

Nematodes are roundworms found in soil and water. Some are free-living while others are parasitic on plants and animals. Nematodes feeding on plant roots beneath the soil surface are especially destructive to crops and other plants. The nematicides listed below are recommended for the control of nematodes attacking most plants and crops.

aldicarb
carbofuran
Dasanit (fensulfuthion)
DBCP
D-D (dichloropropene-dichloro-propane)
Di-Syston (DMTT)
EDB (ethylenedibromide)
methyl bromide ($CH_3Br$)

MIT (Vorlex)
Mocap (prophos)
Nemacide (dichlofenthion)
Nemafos (thionazin)
Penphene (tetrachlorothiophene)
Thimet (phorate)
Vapam (SMDC)
Zinophos (Thionazin)

### Rodenticides

Rodenticides constitute a special group of pesticides that are used chiefly to kill rats and mice, but which also are effective against other animals. Within the rodenticide family there are several subgroups of poisons, such as the *anti-coagulants*, the *quick-acting poisons*, and the *fumigant gases*. Because of their relatively low toxicity and safety to man and pets, the anti-coagulants have become especially popular rodenticides on a worldwide basis. However, they are not always effective and, consequently, the reader may wish to consider the large number of other rodenticides available. The list below includes members of all rodenticide groups. NOTE: see also the *anti-coagulant* group listed separately in this text.

ANTU®
arsenic trioxide
barium carbonate
calcium cyanide
carbon disulfide
chloropicrin
chlorophacinone
Compound 1080
Compound 1081 (flouroacetate)
coumachlor
coumatetralyl
coumafuryl
DDT
dicumarol
diphacinone (Diphacin)
endrin
Fumarin
Fumasol-G
Gophacide

Gopha-Rid
hydrogen cyanide
methyl bromide
norbormide
phosphorus
PID®
Pival
Pivalyn
Pindane
PMP (Valone)
Prolin
Promar
Red Squill
RoZol
sodium flouroacetate (1080®)
strychnine
thallium sulfate
warfarin
zinc phosphide

# PESTICIDE
# MASTER DESCRIPTION
# LIST

This master list of pesticides is provided for quick-reference convenience. It lists, in alphabetical order, the best-known and most commonly used pesticides, and provides a brief, informative description of each. Most readers will find this list quite useful when selecting a pesticide for use against a certain pest (or pests)—or simply for general information about a given pesticide.

The reader will note that, in some cases, two or more names are given for a pesticide. These are the *generic* and *trade* name (or names) of the particular pesticide. In a few cases, a given pesticide is known by only one name. In other cases, the trade-name, indicated by capitalization or the symbol®, is given first, followed by the generic name in parentheses. This is true when both names are included in alphabetical order in the list.

**ABATE®**—see: *Temephos*

**ACEPHATE**—see: *Orthene®*

**ACRYLONITRILE**—highly flammable and toxic. Acrylonitrile (66%) always must be mixed with carbon tetrachloride (34%) before use. This mixture is a good spot fumigant that is effective against nearly all insects and is safe enough to apply to food processing areas. However, it must be handled carefully. The person applying it should wear a respiratory mask and should make sure that people, pets, or wildlife are not exposed to the gas.

**ALDICARB (TEMIK®)**—highly toxic. Temik is used chiefly as a soil treatment pesticide for sugar beet root maggots and as a soil-applied plant systemic insecticide for insect pests on potatoes and some greenhouse plants.

**ALDRIN**—moderately toxic. Aldrin is related chemically to dieldrin and, in fact, frequently converts to dieldrin in soil and plants. In the past, aldrin was used to control a wide variety of insects. In recent years, however, its usage has been restricted by the E.P.A. to that of a soil poison for subterranean (underground) termites.

**ALLETHRIN**—low toxicity. Allethrin is a synthetic (artificial) pyrethrin. Allethrin must be used selectively since it is highly effective against some insects but less effective against others. It is especially useful as a "quick knock-down" insecticide for almost any type of insect or arachnid pest.

**ALTOSID® (METHOPRENE)**—low toxicity. Altosid is a slow-release larvicide that is used by mosquito control districts to kill mosquito larvae in water before they mature to adults. Altosid can be used to kill mosquito and other water-breeding insects around the house, particularly in small ponds, pools, ditches, artificial containers, etc.

**AMAZE®**—highly toxic. Amaze is used principally as a soil insecticide against the corn rootworm.

**AMBUSH®**—see: *Permethrin*

**ANTU®**—low toxicity. ANTU® is a quick-acting poison for Norway rats, but is less effective against roof rats and mice. Rats tend to develop bait-shyness to ANTU® and, thus, it should be applied only at intervals approximately three (3) months apart. No secondary poisoning of dogs, cats, or other animals occurs with ANTU®.

**ARAMITE®**—low toxicity. Discovered in 1951, Aramite is one of the older miticides available today. Aramite contains sulfur which is effective against mites and also acts as a repellent against ticks. Because of its low toxicity to insects, Aramite is used chiefly against mites which are arachnids.

**ARSENIC TRIOXIDE**—highly toxic to man and animals. Known as "white arsenic" and arsenious oxide, arsenic trioxide is used chiefly as a rodenticide. Although it kills slowly, it nevertheless is effective against rats and mice. However, it must be used with extreme caution around children and pets.

**ASPON®**—low toxicity. Aspon finds use as an effective turf and lawn insecticide, especially against the sod webworm and chinch bug.

**AVITROL**—highly toxic. Avitrol is a bird repellent that is used against such nuisance species as pigeons, blackbirds, sparrows, and starlings—on buildings and in roosting/feeding areas. Avitrol is applied as treated grain bait that, when eaten by a few birds, acts to produce distress behavior. This, in turn, repels or frightens away the rest of the flock.

**AZINPHOSMETHYL (GUTHION®)**—highly toxic. Guthion is a general-usage, broad-spectrum insecticide that is effective against a wide variety of insects on fruits, vegetables, flowers, and ornamental plants.

**AZODRIN®**—see: *Monocrotophos*

**BARIUM CARBONATE**—moderately toxic. Barium carbonate is used to kill Norway rats. For good results, however, prebaiting is usually necessary to determine which bait material is most acceptable to the rats. Fresh bait material should be used regularly with this poison.

**BASE OILS**—these are solvents or diluents that are used in many oil-based pesticides. Base oil diluents usually are high-grade (highly refined) kerosenes that are essentially colorless and odorless and that tend to dry "clean," leaving very little if any oil film on treated surfaces.

**BAYGON®**—see: *Propoxur*

**BAYTEX®**—see: *Fenthion*

**BENDIOCARB (FICAM®)**—highly toxic. This is a wide-spectrum insecticide that is especially useful against soil insects. It provides good residual action.

**BENZENE HEXACHLORIDE (BHC)**—moderately toxic. BHC is a contact insecticide with a short residual life. Its strong, musky odor usually limits its usage to outside control problems.

**BHC**—see: *Benzene hexachloride*

**BIDRIN®**—see: *Dicrotophos*

**BINAPACRYL (MOROCIDE®)**—highly toxic. Morocide is used exclusively against mites.

**BIORESMETHRIN**—low toxicity. This is a wide-spectrum insecticide that is especially useful against stored food insects.

**BIRLANE®**—see: *Chlorfenvinphos*

**BLACK FLAG-40®**—see: *Nicotine sulfate*

**BOLSTAR®**—moderately toxic. Bolstar appears to be effective against caterpillars (worm-like larvae that feed on plants).

**BOMYL®**—highly toxic. Bomyl is effective against houseflies.

**BORAX**—low toxicity. Borax is used as a finely powdered dust for roach control. It acts slowly, however, and loses its effectiveness in damp areas.

**BROMOPHOS**—moderately toxic. This is a wide-spectrum insecticide that is especially useful against sucking and biting insects such as those on flowers, plants, and shrubs.

**CALCIUM CYANIDE**—extremely toxic! Like other forms of cyanide (a deadly poison) calcium cyanide should be used *only* by certified pest control operators. This is a powdered or granular material that reacts with moist air to produce hydrocyanic acid gas—a deadly fumigant poison effective against *all* pests.

**CARBARYL (SEVIN®)**—low toxicity. This is a broad-spectrum, general-usage insecticide that is effective against a wide variety of insect species of home, yard, and garden. It also is effective against arachnids (ticks, mites, spiders, etc.). Sevin is effective against many insect species that are resistant to other poisons. However, it is *not* effective against the following insects: houseflies, carpet beetles, and termites. Incidentally, Sevin is highly toxic to honeybees. Generally, Sevin can be used with effectiveness against all types of ticks in yards, houses, and dogpens. It also is effective against evergreen bagworms.

**CARBOFURAN (FURADAN®)**—highly toxic. Like most soil-applied systemics, Furadan is highly toxic to man and animals. It is used effectively against: corn rootworms, European corn borers, sugarbeet root maggots, alfalfa weevils, grasshoppers, and potato insects.

**CARBON DISULPHIDE**—extremely flammable and moderately toxic rodenticide. Carbon disulphide displays a stringent, irritating odor. Because of its flammability and odor, carbon disulphide *never* should be used in, under, or near buildings. Rather, it should be used only in rat or mole burrows that are out in the open and located away from combustible things.

**CARBON MONOXIDE**—highly toxic. This very toxic gas is a primary component resulting from the combustion (burning) of petroleum (carbon) products. Carbon monoxide (CO) is particularly prevalent in automobile exhaust emissions, and in the exhaust from heaters and other burning materials.

In some cases, the CO emitted in your automobile exhaust can be used safely, effectively, and cheaply to fumigate certain pests. For example, you can connect a hose to your car's exhaust pipe and then place the open end of the hose into the openings of underground *burrows* to fumigate rats, mice, snakes, moles, gophers, and many insects and other pests. This method id particularly effective and cheap—and even spares you the task of disposing of the dead carcasses which already are buried.

Carbon monoxide also has been used to fumigate bats in attics and between the walls of houses. However, this method is *very dangerous*. It must be employed *only* with the very strictest supervision and caution. *All* human beings and pets MUST be removed from the building BEFORE it is fumigated. NOTE: CO is heavier than air and will settle downward from the attic, replacing the air and permeating all rooms within the house. Therefore, following the fumigation, you must open all doors and windows and ventilate the house or building thoroughly before anyone occupies it again. Remember: CO is odorless, colorless, and tasteless. You cannot detect its presence. Symptoms of carbon monoxide poisoning are: frontal headache, drowsiness, confusion, etc.

**CARBON TETRACHLORIDE**—highly toxic. Carbon tetrachloride is a clear liquid that produces a volatile toxic fumigant gas. Carbon "tet" can be used alone or with two other synergistic fumigant gases—ethylene dibromide and ethylene dichloride. As a fumigant, carbon tet is effective against *all* pests, but its high toxicity makes it very hazardous to apply. It *must* be applied using a gas mask or respirator.

**CARBOPHENOTHION (TRITHION®)**—highly toxic. Trithion is a broad spectrum insecticide and miticide that is especially useful on fruits, vegetables, flowers, and plants.

**CARBOXIDE**—see: *Ethylene oxide*

**CARZOL**—see: *Formetanate*

**CHLORBENSIDE (MITOX®)**—low toxicity. This chemical is effective against mites on fruits, flowers, and plants.

**CHLORDANE**—low toxicity. Like most other chlorinated hydrocarbons of the DDT family, chlordane usage has been restricted by the E.P.A. because of its chemical stability and long-term residual power that make it a threat to man and animals through the food and water chains. Chlordane now is used mainly to control subterranean (underground) termites beneath buildings.

**CHLORDECONE (KEPONE®)**—highly toxic. Although Kepone usage has been restricted by the E.P.A., it is very effective in bait form to control cockroaches and ants. These baits, however, must be placed out of reach of children and pets.

**CHLORDIMEFORM (GALECRON®; FUNDAL®)**—highly toxic. These poisons are especially useful against the eggs and larvae (caterpillars) of several moths that attack crops, as well as against most stages of ticks and mites and many other insect species.

**CHLORFENVINPHOS (COMPOUND 4072®; BIRLANE®)**—highly toxic. Compound 4072 is used princially to kill flies.

**CHLORMEPHOS**—highly toxic. This soil insecticide is very effective against wireworms, millipedes, symphilids, and many other pests.

**CHLOROBENZILATE**—low toxicity. This chemical is used chiefly against mites.

**CHLOROPHACINONE (ROZOL®)**—low toxicity. RoZol is an anti-coagulant cumulative poison that causes slow internal bleeding. It is used against Norway rats and house mice. Like most other anti-coagulant baits, RoZol requires repeated feedings over a period of days for good results. Note: RoZol can be absorbed through the *skin* and, thus, should be handled with rubber gloves.

**CHLOROPICRIN (TEAR GAS)**—fairly toxic. "Tear gas" should be handled with caution. It is a good spot fumigant against a variety of pests, and is used as a warning gas for other more toxic, but less detectable fumigants.

**CHLORPYRIFOS (DURSBAN®; LORSBAN®)**—moderately toxic. Dursban is especially effective against cockroaches and many other insects as well. Although it acts somewhat slowly, Dursban provides good residual action, particularly on smooth and non-porous surfaces such as stainless steel, linoleum, tile, etc.

**CIODRIN®**—see: *Crotoxyphos*

**COMPOUND 4072®**—see: *Chlorfenvinphos*

**CO-RAL®**—see: *Coumaphos*

**COUMAPHOS (CO-RAL®)**—low toxicity. Coumaphos is an animal systemic (given internally) by veterinarians to control fleas on dogs over two months old, but not on cats. Co-Ral also is used to control grubs, lice, and flies on livestock.

**COUNTER®**—see: *Terbufos*

**CROTOXYPHOS (CIODRIN®)**—moderately toxic. Ciodrin is used principally to control lice and flies on livestock.

**CRUFOMATE (RUELENE®)**—low toxicity. Ruelene is used chiefly as an animal systemic (given internally) pesticide to control ectoparasites (pests on outside of body) such as: fleas, ticks, lice, mites, etc.

**CYGON®**—see: *Dimethoate*

**CYHEXATIN (PLICTRAN®)**—moderately toxic. Plictran is a relatively new pesticide that is selectively useful against mites. Plictran also doubles as a fungicide that is effective against some molds, mildew, and other household fungi.

**DASANIT®**—see: *Fensulfothion*

**D-CON®**—see: *Warfarin*

**DDD**—see: *TDE*

**DDT**—moderately toxic. Worldwide, DDT is one of the best known and, formerly, most widely-used of all pesticides. Because of its long residual life in soil and water food chains, however, DDT usage has been restricted greatly. Nevertheless, DDT is still highly effective against a wide variety of insect, arthropod, and even vertebrate pests.

**DDVP**—see: *Dichlorvos*

**DEFEND®**—see: *Dimethoate*

**DELNAV®**—see: *Dioxathion*

**DEMETON (SYSTOX®)**—highly toxic. Systox is a foliar (plant) systemic pesticide used to control sap-sucking insects on field, fruit, and vegetable crops and on ornamental flowers and plants.

**DIAZINON (SPECTRACIDE®)**—moderately toxic. Diazinon is a broad spectrum, general usage insecticide that is effective against a wide range of insect pests, including: household insects, field insects, fruit, yard, vegetable, and ornamental plant insects. Diazinon also is effective against houseflies and their larvae (maggots) and against ticks and other arachnids. However, a diazinon emulsion should be used soon after it is made up because its effectiveness is lost when left standing for too long.

**DIBROM®**—see: *Naled*

**DICHLORVOS (VAPONA®; DDVP)**—highly toxic. This is a good general-use insecticide for use in enclosed areas. It is particularly useful in dog-and-cat flea/tick collars and in vapor-release resin strips for the control of flying insects indoors. Dichlorvos also can be used effectively to control pantry, stored food, carpet, and fabric pests, as well as bedbugs and cockroaches.

**DICOFOL**—see: *Kelthane®*

**DICROTOPHOS (BIDRIN®)**—highly toxic. Dicrotophos is an organophosphate plant systemic insecticide that kills sap-sucking insects that feed on plant juices. Normally, caterpillars and other chewing pests are not controlled by dicrotophos because they eat the plant tissues rather than suck the sap or juices.

**DIESEL FUEL/FUEL OIL/MOTOR OIL**—Although these common petroleum products generally are not considered pesticides, they can in fact be used effectively to kill or repel a variety of pests in certain cases.

For example, diesel fuel has been used for many years, and continues to be used, by health departments and mosquito abatement districts to control mosquitoes. Sprayed as a thin film on water in ditches, potholes, swamps, storm drains, containers, discarded tires, etc., diesel or motor oil causes suffocation of mosquito larvae and other water-breeding insect larvae, thus killing them before they mature to flying adults. This control method is both economical and very effective. Thus, either diesel fuel or crankcase motor oil can be used around the home to effectively control insects, especially mosquitoes and midges, that breed in water-filled containers such as fishponds, ditches, flower pots, discarded tires, and other water-holders.

Another effective use of diesel, fuel oil, or motor oil is to control, or repel, the migration of ticks and other crawling arthropods into your yard or other confined area. A 3–4 inch wide band or barrier of diesel, fuel, or motor oil, applied continuously around the perimeter of the area to be protected, is, in most cases, quite effective in repelling ticks and other crawling invaders. Likewise, such a barrier applied around the perimeter of your house (or doghouse) may be effective in preventing the entrance of ticks and other crawling pests into these structures.

Still another little-known use of diesel fuel, but *not* motor oil, is in controlling subterranean termites beneath buildings. It must be emphasized, however, that diesel does not normally kill the termites, at least not in great numbers, but rather, repels them. In order to repel effectively, however, the soil beneath the building must be *soaked* and saturated with diesel fuel. Note: while this method is usually effective and certainly less expensive than the standard chemical treatment, it cannot be guaranteed— simply because the termites are repelled rather than killed.

**DIMECRON®**—see: *Phosphamidon*

**DIMETHOATE (CYGON®; DEFEND®; REBELATE®; DIMEX®)** —moderately toxic. Dimethoate is used chiefly as a residual spray to control flies and fly larvae (maggots), and as a soil-applied plant systemic to control a broad spectrum of plant-feeding insects on fruit, field, vegetable, and ornamental crops.

**DIMETHRIN**—low toxicity. This is a *synthetic* pyrethrin pesticide that is very similar to pyrethrin in the natural form.

**DIMETILAN**—highly toxic. Like Vapona and DDVP, dimetilan kills insects with its toxic *vapor* that is emitted slowly from plastic or resin strips in which it usually is impregnated. Dimetilan is especially useful against flies and other flying insects inside buildings.

**DIMEX®**—see: *Dimethoate*

**DINOCAP (KARATHANE®)**—low toxicity. Karathane is useful for controlling powdery mildew fungi.

**DIOXATHION (DELNAV®)**—moderately toxic. Delnav is useful as a spot-treatment poison for brown dog ticks and for insects and mites on fruit trees and livestock.

**DIPHACIN®**—see: *Diphacinone*

**DIPHACINONE (DIPHACIN®; PID®)**—highly toxic. This is an anti-coagulant rodenticide that is effective against all species of rats and mice. Like most other anti-coagulants (which work slowly by causing gradual internal bleeding), Diphacin is cumulative and thus requires several feedings over a period of days for satisfactory results. Neither rats nor mice develop bait-shyness to Diphacin, which can be used either as a dry powder or a liquid bait.

**DIPTEREX®**—see: *Trichlorfon*

**DISULFOTON (DI-SYSTON®)**—highly toxic. Di-Syston is a soil-applied plant systemic poison useful for controlling sap-sucking insects on potatoes, vegetables, ornamental plants, and flowers. Di-Syston also is effective against aphids on some field crops and against the sugarbeet root maggot.

**DI-SYSTON®**—see: *Disulfoton*

**DRC-139®**—see: *Starlicide*

**DURSBAN®**—see: *Chlorpyrifos*

**DYFONATE® (FONOFOS)**—highly toxic. This is a soil insecticide that is effective against corn rootworms, wireworms, sugarbeet root maggots, and other soil pests.

**DYLOX®**—see: *Trichlorfon*

**EDB**—see: *Ethylene dibromide*

**EDC**—see: *Ethylene dichloride*

**ENDOSULFAN (THIODAN®)**—moderately toxic. Thiodan is a broad-spectrum insecticide that is especially useful for controlling both insects and mites on vegetables, fruit, and ornamental plants. Thiodan also is useful against greenhouse pests.

**ENDRIN**—highly toxic. Endrin is a close chemical relative of aldrin and dieldrin and like its chemical cousins, endrin usage, too, has been restricted somewhat. Besides its general insecticidal effectiveness, endrin also is useful for controlling pigeons and other perch-resting birds inside buildings.

**ENSTAR®**—see: *Kinoprene*

**ENTEX®**—see: *Fenthion*

**EPN (PHENYLPHOSPHOROTHIOATE)**—highly toxic. EPN is useful for controlling European corn borers, fruit pests, and mosquitoes.

**ETHION**—highly toxic. Ethion is particularly useful for controlling mites on fruits, vegetable plants, flowers, ornamental plants, and also for controlling scales and onion maggots.

**ETHOPROP (MOCAP®)**—highly toxic. Mocap is especially useful against corn rootworms and wire worms.

**ETHYLENE DIBROMIDE (EDB)**—moderately toxic. EDB is an effective nematocide. It controls nematodes and their eggs, as well as soil insects, when injected into the soil of seed beds, flower beds, potting soil, and greenhouse soils. EDB, when mixed with ethylene dichloride and carbon tetrachloride, is an effective grain fumigant against many grain-feeding pests.

**ETHYLENE DICHLORIDE (EDC)**—moderately toxic. EDC is a liquid poison, the gas (vapor) of which is a good fumigant when mixed with EDB and carbon tetrachloride.

**ETHYLENE OXIDE (CARBOXIDE)**—moderately toxic, highly flammable, and explosive. Ethylene oxide (EO) can be bought mixed 1 part EO to 9 parts carbon dioxide, which reduces the flammability danger. EO is useful for fumigating grain and stored food products. EO also has bactericidal properties, thus making it useful for *disinfecting* individual items as well as entire buildings.

**FAMPHUR (WARBEX®)**—highly toxic. This pesticide is useful for controlling cattle grubs and cattle lice.

**FENSULFOTHION (DASANIT®)**—highly toxic. Dasanit is especially useful for controlling corn rootworms, onion maggots, and other root maggots.

**FENTHION (BAYTEX®; ENTEX®)**—moderately toxic. Fenthion is sold in two commercial forms: *Baytex* for outside use and *Entex* for indoor use. Fenthion is especially useful for controlling flies, mosquitoes, and other flying pests, including most common household insects. Additionally, fenthion is effective against pigeons, starlings, and sparrows inside buildings when used on perches frequented by these pests.

**FENSON**—low toxicity. Fenson is useful for controlling mites on fruit crops.

**FENVALERATE (PYDRIN®)**—low toxicity. Fenvalerate is a pyrethrin-derivative that is effective against a wide range of insects, especially caterpillars. Pydrin can be used on fruits, vegetables, flowers, and other plants.

**FICAM®**—see: *Bendiocarb*

**FLUOROACETAMIDE (1081®)**—*extemely toxic!* 1081® is very similar to sodium fluoroacetate (1080®). Both chemicals are *extremely* poisonous and are used chiefly to kill rats and mice. These poisons act quickly, even in small doses, and are very hazardous to use. NOTE. There is *no known antidote* to either 1080® or 1081®. Thus, it is absolutely imperative that you place them out of reach of children, pets, and non-target animals. Locked bait boxes should be used to prevent people or animals from accidentally contacting these poison baits. A rat or mouse carcass killed by either 1080® or 1081® contains enough poison to kill a cat or dog which eats the rodent carcass. Nevertheless, 1081® is very effective as a rodenticide and also against most other vertebrate or animal pests.

**FONOFOS**—see: *Dyfonate®*

**FORMETANATE (CARZOL)**—low toxicity. This chemical is similar to Sevin and Lannate.

**FUMARIN®**—low toxicity. This is an anti-coagulant rodenticide that is effective against all species of rats and mice. Like most other anti-coagulants, Fumarin is a cumulative poison that requires a number of feedings over a period of days for good kill results. Neither rats nor mice develop bait-shyness to Fumarin (the dry bait form) nor to Fumasol (the liquid bait form).

**FUNDAL®**—see: *Chlordimeform*

**FURADAN®**—see: *Carbofuran*

**GALECRON®**—see: *Chlordimeform*

**GARDONA®**—see: *Stirofos*

**GENITE®**—moderately toxic. Genite is chemically similar to Aramite, ovex, and Tedion. It is a good general-purpose miticide (acaricide) that also is effective against some insects.

**GUTHION®**—see: *Azinphosmethyl*

**HCN**—see: *Hydrogen cyanide*

**HEPTACHLOR**—moderately toxic. Heptachlor is chemically related to DDT and the other chlorinated hydrocarbon pesticides. And heptachlor usage, too, has been restricted, essentially, to subterranean termite control beneath buildings. Heptachlor is more volatile and, thus, emits greater vapor toxicity than the other members of this chemical group.

**HYDROGEN CYANIDE (HCN)**—*extremely* toxic gas! This is, in fact, the gas that is used most often to execute condemned prisoners in penitentiary gas chambers. Even a few inhaled breaths of this gas is sufficient to render a person or animal unconscious. Death follows quickly from respiratory muscle paralysis. HCN is used *only* by certified pest control operators who employ it most often to fumigate ground-burrows for such pests as rats, snakes, moles, and gophers.

**IMIDAN®**—see: *Phosmet*

**KARATHANE®**—see: *Dinocap*

**KELTHANE® (DICOFOL)**—low toxicity. While primarily an acaricide, Kelthane also is effective against many insects. Its chief use, however, is in controlling a wide variety of mites that infest buildings, trees, plants, shrubs, flowers, and lawns.

**KEPONE®**—see: *Chlordecone*

**KINOPRENE (ENSTAR®)**—low toxicity. Kinoprene is useful for controlling whiteflies and aphids on flowers and other plants in greenhouses, and for this type of pest control problem outside greenhouses as well.

**KORLAN®**—see: *Ronnel*

**LANNATE®**—see: *Methomyl*

**LETHANE® (THANITE)**—low toxicity. Lethane gives quick knockdown of most flying insects, and can be used both inside the house and outside.

**LINDANE**—moderately toxic. Lindane quickly paralyzes most insect species. Unlike most other insecticides, lindane is stable to heat, and thus can be used in and around heat radiators such as heaters, radiators, stoves, and boilers, without losing its effectiveness. However, lindane is more volatile than DDT and, thus, provides less residual effectiveness.

**LORSBAN®**—see: *Chlorpyrifos*

**MALATHION**—very low toxicity. Malathion is a very good general-purpose, broad-spectrum insecticide for all-around household use. It is safe enough for inside use, and is effective against a wide range of insects, mites, and aphids found inside the house, in gardens, on fruit and vegetables, and on trees and shrubs. Malathion also is effective against flies, mosquitoes, and other flying insects. Although it gives good residual action, malathion tends to break down rather quickly on contact with water and when left standing in a container for very long.

**MESUROL®**—see: *Methiocarb*

**META-SYSTOX-R®**—see: *Oxydemetonmethyl*

**METHAMIDOPHOS (MONITOR®)**—highly toxic. Monitor is used to control insects on potatoes and some other vegetables. It is especially useful against aphids and loopers.

**METHIDATHION (SUPRACIDE®)**—highly toxic. Supracide is used principally to control insects on alfalfa and sunflowers.

**METHIOCARB (MESUROL®)**—moderately toxic. Mesurol is useful against slugs, snails, sowbugs, pillbugs, and some fruit insects. Mesurol also is a bird repellent.

**METHOMYL (LANNATE®)**—moderately toxic. Especially good for controlling worms and caterpillars on vegetables.

**METHOPRENE**—see: *Altosid®*

**METHOXYCHLOR**—low toxicity. Methoxychlor is chemically related to DDT, but is much less stable and tends to break down within a few weeks after application. Because of its low toxicity and broad-spectrum effectiveness, methoxychlor is useful against a wide range of indoor and outdoor insects, including: pests of vegetables, fruits, shrubs, flowers, and ornamental plants. Methoxychlor also is useful against hornflies on cattle.

**METHYL BROMIDE**—highly toxic. This liquid vaporizes into a gas at temperatures above 37 degrees F. Its good penetrating power makes MB an effective fumigant against a number of pests. It is non-flammable. However, MB is odorless and, thus, should be mixed with an odorous gas before being used.

**METHYL PARATHION**—extremely toxic. This chemical should *not* be used by the do-it-yourselfer because of the great hazard it presents to people and animals.

**MEVINPHOS (PHOSDRIN®)**—highly toxic. Phosdrin is used chiefly on a large commercial scale to control pests of fruit and vegetable crops.

**MEXACARBATE (ZECTRAN®)**—highly toxic. Zectran is used to control mites, snails, sowbugs, and pillbugs.

**MITOX®**—see: *Chlorbenside*

**MOCAP®**—see: *Ethoprop*

**MONITOR®**—see: *Methamidophos*

**MONOCROTOPHOS (AZODRIN®)**—highly toxic. Azodrin is used chiefly to control potato insects.

**MORESTAN®**—see: *Oxythioquinox*

**MOROCIDE®**—see: *Binapacryl*

**MOTHBALLS/FLAKES**—see: *Naphthalene*

**NALED (DIBROM®)**—moderately toxic. Naled is especially useful for controlling flying insects both indoors and outside. It also is effective against a wide variety of fruit and vegetable insects. Naled breaks down fairly quickly in the environment.

**NAPHTHALENE (MOTHBALLS, MOTHFLAKES)**—very low toxicity. Naphthalene is used commonly in the form of mothballs and mothflakes for protecting clothing, carpets, woolens, etc. against moths and carpet beetles. Mothballs and flakes also are effective in repelling bats in attics and other roosting areas.

**NICOTINE SULFATE (BLACK FLAG-40®)**—moderately toxic. Nicotine sulfate is useful for controlling aphids and some other insects in home gardens and greenhouses. It also can be used in a water-mixture to kill snakes. One (1) part of 40% nicotine sulfate (Black Flag-40) is added to 250 parts water and placed in a *shallow* pan which is covered with a screen wire or other covering. The covering is stapled to four wooden pegs, one on each corner of the pan. This leaves just enough room between the wire and the pan (approximately 1½ inches) for snakes to enter and drink the poison, but prevents the entrance of other non-target animals.

**NORBORMIDE**—low toxicity. Norbormide is a rodenticide that is specifically for Norway rats. It is not effective against roof rats nor mice.

**OMITE®**—low toxicity. Omite is specifically for mites, particularly on fruits.

**ORNITROL®**—highly toxic. Ornitrol is a chemosterilant (causes sexual sterilization) that is specific for pigeons when fed to them on corn grains.

**ORTHENE® (ACEPHATE)**—low toxicity. Orthene is particularly useful for controlling aphids, grasshoppers, loopers, thrips, and caterpillars on flowers and other types of plants.

**OVEX (OVOTRAN®)**—low toxicity. Ovex is useful against mites on fruit trees and plants.

**OVOTRAN®**—see: *Ovex*

**OXAMYL (VYDATE®)**—highly toxic. Vydate seems effective in controlling insects on most flowering plants. It also is an effective nematocide.

**OXYDEMETONMETHYL (META-SYSTOX-R®)**—highly toxic. This chemical is useful for controlling insects in home gardens, aphids, mites and leaf hoppers, plus insects on flowers and ornamental plants.

**OXYTHIOQUINOX (MORESTAN®)**—low toxicity. Morestan is useful for controlling mites on fruits and flowers.

**PARADICHLOROBENZENE (PDB)**—low toxicity. PDB is very similar to naphthalene and, in fact, often is used to replace naphthalene in mothballs and flakes. The odor of PDB is less persistent than that of naphthalene and, for this reason, PDB is more satisfactory for treating clothing—particularly woolens. PDB, like naphthalene, also can be used to repel bats.

**PARATHION**—extremely toxic. Parathion is much too toxic for use by the general homeowner. Thus, it is restricted to use only by certified pest control operators.

**PCP**—see: *Pentachlorophenol*

**PDB**—see: *Paradichlorobenzene*

**PENTAC®**—low toxicity. Pentac is effective against mites on trees and shrubs, flowers, and greenhouse crops.

**PENTACHLOROPHENOL (PCP)**—moderately toxic. PCP is used mainly as a wood preservative, but also is effective against drywood termites and fungi. PCP is extremely irritating to the eyes and skin and must be handled with great care. As with most pesticides, PCP handling and application should be done wearing rubber gloves.

**PERMETHRIN (AMBUSH®; POUNCE®)**—low toxicity. This is a synthetic pyrethrin product that is effective against a wide range of insects, especially caterpillars (worm-like larvae of various insects that feed on trees, plants, and crops).

**PERTHANE®**—moderately toxic. Perthane is related chemically and is similar to DDT.

**PHENYLPHOSPHOROTHIOATE**—see: *EPN*

**PHORATE (THIMET®)**—highly toxic. This is a soil-applied plant systemic that is effective against corn rootworms, potato and vegetable insects.

**PHOSDRIN®**—see: *Mevinphos*

**PHOSMET (IMIDAN®; PROLATE®)**—moderately toxic. Phosmet is effective against alfalfa weevils and a wide variety of fruit and shrub pests. The Prolate form is useful against livestock pests.

**PHOSPHAMIDON (DIMECRON®)**—highly toxic. Dimecron is used on a commercial scale against pests of fruit and vegetable crops—particularly aphids, mites, and leafhoppers.

**PHOSPHINE (PHOSTOXIN®)**—moderately toxic. Phosphine granules or tablets, when scattered through stored grain, react with moist air to release phosphide gas which acts as a fumigant against grain insects, leaving behind only a harmless powdered residue.

**PHOSPHORUS**—extremely toxic. Phosphorus is too dangerous for use by the general homeowner, and is limited to employment by certified pest control operators who use it primarily to control difficult infestations of rats and cockroaches.

**PHOSTOXIN®**—see: *Phosphine*

**PID®**—see: *Diphacinone*

**PIRIMICARB (PIRIMOR®)**—moderately toxic. Pirimor is particularly useful against aphids on greenhouse plants and on potatoes.

**PIRIMOR®**—see: *Pirimicarb*

**PIVAL®**—moderately toxic. Pival is an anti-coagulant rodenticide that is effective against both rats and mice, but which requires the usual several feedings over a period of days for good kill results. Pival (dry bait form) or Pivalyn (liquid bait form) can be used repeatedly since rodents do not develop bait-shyness to either form.

**PLICTRAN® (CYHEXATIN)**—moderately toxic. Plictran is selectively effective against mites and, possibly, ticks also. Additionally, Plictran can be used to control fungi.

**PMP (VALONE®)**—low toxicity. Like most other anti-coagulant rodenticides, Valone requires several feedings over a number of days for good kill results. It is effective against all species of rats and mice, none of which develop bait-shyness to it.

**POUNCE®**—see: *Permethrin*

**PROLATE®**—see: *Phosmet*

**PROPOXUR (BAYGON®)**—highly toxic. Baygon is very effective against cockroaches, ticks, and other difficult insect and arachnid species, including household, lawn, turf, and garden pests. Its quick flushing action and long residual life make Baygon a very useful general-purpose insecticide.

**PROXOL®**—see: *Trichlorfon*

**PYDRIN®**—see: *Fenvalerate*

**PYRETHRINS**—see: *Pyrethrum*

**PYRETHRUM (PYRETHRINS)**—low toxicity. Pyrethrum is a natural botanical insecticide that is made from an extract taken from the chrysanthemum plant. Pyrethrum provides quick knock-down and quick flushing action, but it does not always kill the insects against which it is used. For this reason, pyrethrum usually is mixed with the synergist, piperonyl butoxide, or other insecticides. Pyrethrum is effective against household, stored food, and garden insects.

**RABON®**—see: *Stirofos*

**REBELATE®**—see: *Dimethoate*

**RED SQUILL®**—low toxicity. Red Squill is a rodenticide that is more effective against Norway rats than against roof rats and house mice. Also, rats tend to develop bait-shyness to Red Squill and, thus, it is less useful than some other rodenticides now available.

**RESMETHRIN (SBP-1382®)**—low toxicity. Resmethrin is a synthetic pyrethrum that is effective against whiteflies, fungus gnats, and flower thrips in greenhouses, and against all types of flying and crawling insects indoors and outdoors.

**ROTHANE®**—see: *TDE*

**RONNEL (KORLAN®; TROLENE®)**—low toxicity. Ronnel is especially useful for controlling flies, resistant roaches, and ectoparasites (fleas, lice, mites, etc.) of animals.

**ROTENONE**—moderately toxic. Rotenone is a botanical pesticide that is especially good for controlling garden and vegetable insects. It is extremely toxic to, and provides rapid kill of, fish.

**ROZOL®**—see: *Chlorophacinone*

**RUELENE®**—see: *Crufomate*

**RYANIA**—low toxicity. Ryania is a natural (botanical) insecticide derived from a plant. It is particularly useful against European corn borers and codling moths.

**SABADILLA**—low toxicity. Sabadilla is especially useful against squash bugs and other plant bugs.

**SBP-1382®**—see: *Resmethrin*

**SEVIN®**—see: *Carbaryl*

**SILICA AEROGEL**—non-toxic to humans. Silica aerogel is a desiccant insecticide. That is, it kills by disrupting or dissolving the waxy outer coat (epicuticle) of the insect body, thus inducing excessive water and body-fluid loss and leading to death from dehydration. Silica aerogels are somewhat difficult to apply, however, because they are light and fluffy and, thus, difficult to keep confined to the treated area. Better results are obtained when the silica aerogel is mixed with another insecticide such as pyrethrin. This increases the mass and serves to keep the chemical in place where it is applied. Silica aerogel used in this manner is useful for cockroach control on a long-term basis—particularly between walls and floors and other hard-to-reach places.

**SODIUM FLUORIDE**—extremely toxic. This white powder poison is required by law to be mixed with a blue or green powder so that it is not mistaken for food materials such as flour, sugar, salt, etc. Like most powdered insecticides, sodium fluoride must be kept *dry* for full effectiveness. It is useful for controlling cockroaches and many other insects.

**SODIUM FLUOROACETATE (1080®)**—extremely toxic! This unusually poisonous rodenticide is effective against both rats and mice, but is *very hazardous* to handle and use. A very small amount of 1080® is fatal to both humans and animals. For instance, a rodent carcass killed by 1080® or 1081® contains enough poison to, in turn, kill a cat or dog which eats it. NOTE: There is no known antidote to either 1080® or 1081®. Thus, both must be handled with extreme caution and applied only in *locked* bait boxes with openings no larger than $2\frac{1}{2} \times 2\frac{1}{2}$ inches.

**SPECTRACIDE®**—see: *Diazinon*

**STARLICIDE (DRC-1339®)**—moderately toxic. This is a bird repellent that is useful in pellet form to control starlings, in particular, and some other pest species as well.

**STIROFOS (GARDONA®; RABON®)**—low toxicity. Gardona is a general-purpose home-safe insecticide that is effective against a wide variety of household and garden insects. The Rabon form is especially effective against flies.

**STROBANE®**—see: *Toxaphene*

**STRYCHNINE**—highly toxic. Strychnine is a very toxic substance that long has been used to poison *animal* pests. It must be handled with great care and placed *only* where the intended pests will contact it. Strychnine is effective against rats, mice, birds, snakes, moles, gophers, squirrels, foxes, coyotes, and most other higher animals.

**SULFURYL FLUORIDE (VIKANE®)**—moderately toxic. Vikane is an odorless, non-flammable fumigant gas that is effective against a number of different insects. It is especially useful against drywood termites because its high penetrating power carries it deep into infested wood.

**SUPRACIDE®**—see: *Methidathion*

**SYSTOX®**—see: *Demeton*

**TDE (DDD; ROTHANE®)**—low toxicity. Chemically, TDE (DDD) is very similar to DDT and is effective against a wide range of insects.

**TEAR GAS**—see: *Chloropicrin*

**TEDION®**—see: *Tetradifon*

**TEMIK®**—see: *Aldicarb*

**TEMEPHOS (ABATE®)**—low toxicity to mammals, including man, and to birds and fish. Abate is used chiefly to control mosquitoes by killing their larvae or "wigglers" in water before they mature to flying adults. Abate also can be used to control the immature forms of other water-breeding insects such as various midges and flies.

**1080®**—see: *Sodium fluoroacetate*

**1081®**—see: *Fluoroacetamide*

**TEPP (TETRAETHYL PYROPHOSPHATE)**—extremely toxic. No longer manufactured, TEPP formerly was used against greenhouse insects.

**TERBUFOS (COUNTER®)**—highly toxic. Counter is used effectively against corn rootworms, wireworms on corn, and sugarbeet root maggots.

**TETRADIFON (TEDION®)**—low toxicity. Tedion is useful against mites on fruit crops, flowers, and ornamental plants.

**TETRAETHYL PYROPHOSPHATE**—see: *TEPP*

**THANITE**—see: *Lethane®*

**THIMET®**—see: *Phorate*

**THIODAN®**—see: *Endosulfan*

**TOXAPHENE (STROBANE®)**—moderately toxic. Toxaphene is effective against cutworms, grasshoppers, army worms on crops, livestock pests, and ticks.

**TRICHLORFON (DIPTEREX®; DYLOX®; PROXOL®)**—moderately toxic. Dipterex is especially useful against cockroaches and, in sugarbait form, against flies.

**TRITHION®**—see: *Carbophenothion*

**TROLENE®**—see: *Ronnel*

**VALONE®**—see: *PMP*

**VAPONA®**—see: *Dichlorvos*

**VENDEX®**—low toxicity. Vendex is useful against mites on flowers, ornamental plants, fruits, and in greenhouses.

**VIKANE®**—see: *Sulfuryl fluoride*

**VYDATE®**—see: *Oxamyl*

**WARBEX®**—see: *Famphur*

**WARFARIN (D-CON® and other name-brand products)**—low toxicity. Warfarin was the first commercial anti-coagulant rodenticide, having first appeared in the early 1950s. It is a cumulative poison that is effective against all species of rats and mice, but requires several feedings over a period of days for satisfactory results. *Warfacide* is the liquid bait form of this poison.

**ZECTRAN®**—see: *Mexacarbate*

**ZINC PHOSPHIDE**—extremely toxic. This is an effective poison against all species of rats and mice and many other animals as well. However, it must be handled and applied with great caution because of the acute danger it presents to people and animals.

# BROAD-SPECTRUM
# OR GENERAL-USAGE
# PESTICIDES

The following list is provided for your convenience in determining quickly which chemicals are considered effective all-around, general-purpose pesticides against a wide variety of insects and other pests. It should be noted, however, that none of these chemicals can be guaranteed to work equally well against all pests. The *general-purpose* classification given here simply means that each of these pesticides has been found effective against a number and variety of different pest species. Nevertheless, readers will in most cases find this group of pesticides to prove of greater over-all usefulness because of their wide-range effectiveness against many different pest species.

**ACRYLONITRILE**—fumigant gas effective against nearly all pests.

**ALLETHRIN**—low toxicity general-purpose pesticide noted for its quick knock-down and rapid insect flushing action. Allethrin is highly effective against some insects, but less effective against others.

**AZINPHOSMETHYL (GUTHION®)**—broad-spectrum pesticide effective against a wide variety of insects found on fruits, vegetables, flowers, and ornamental plants.

**BENDIOCARB (FICAM®)**—wide-spectrum pesticide effective against many different soil insects and pests.

**BROMOPHOS**—wide-spectrum effectiveness against most sapsucking and biting insects.

**CARBARYL (SEVIN®)**—highly useful general-purpose pesticide that is effective against many different insect and arachnid pests. Sevin can be used both indoors and outdoors.

**CARBON TETRACHLORIDE**—highly toxic fumigant gas that is efffective against *all* pests. Carbon "tet" is hazardous to use, however, and the person applying it must wear a gas mask or respirator.

**CARBOPHENOTHION (TRITHION®)**—though highly toxic, Trithion shows broad-spectrum effectiveness against many different insect and mite species on fruits, vegetables, flowers, plants, and yards.

**CHLOROPICRIN (TEAR GAS)**—a good spot fumigant that is effective against a variety of different pests.

**CHLORPYRIFOS (DURSBAN®; LORSBAN®)**—though somewhat toxic, Dursban is especially effective against cockroaches and many other insects as well, both indoors and outdoors. Dursban acts somewhat slowly but provides good residual action, particularly on smooth and nonporous surfaces such as stainless steel, linoleum, tile, etc.

**DDT**—though its usage is now restricted, DDT still remains one of the very best and most effective of all general-purpose pesticides.

**DIAZINON (SPECTRACIDE®)**—a very good general-purpose pesticide that is effective against many different insect and arachnid pests, both indoors and outdoors. Diazinon is slightly more toxic than either Sevin or malathion.

**DICHLORVOS (DDVP; VAPONA®)**—though highly toxic to humans, this pesticide emits a vapor killing power that is effective against a variety of flying and household insects.

**ENDOSULFAN (THIODAN®)**—wide-spectrum effectiveness against a variety of insects and mites found on vegetables, fruits, flowers, yards, and in greenhouses.

**FENTHION (BAYTEX®; ENTEX®)**—Fenthion is a useful general-purpose pesticide against flies, mosquitoes, and other flying insects, and against most household insects. Also, these pesticides are effective against birds when applied to roosting areas and perches, and they are effective against bats.

**FENVALERATE (PYDRIN®)**—low toxicity and wide-range effectiveness make this a good general-purpose outdoor pesticide. It is equally useful against caterpillars and many other insects and mites found on fruits, vegetables, flowers, yards, and in greenhouses.

**LINDANE**—resistant to heat breakdown and, therefore, effective around heat generators such as stoves, furnaces, boilers, etc. Lindane is effective against a wide range of insects and arachnids.

**MALATHION**—this is a very good general-purpose pesticide that is excellent for all-around home pest control problems. Malathion shows very low toxicity and can be used effectively both indoors and outdoors.

**METHOXYCHLOR**—this is another very good general-purpose pesticide that is very useful to the do-it-yourself pest controller. Methoxychlor toxicity is very low and it is effective against a wide range of indoor and outdoor pests.

**NALED (DIBROM®)**—especially effective against flying insects, both indoors and outside. Dibrom also is effective against a wide variety of fruit, vegetable, flower, greenhouse, and yard insect and arachnid pests.

**PERMETHRIN (AMBUSH®; POUNCE®)**—effective against a wide range of insects of yard, turf, garden, crop, and greenhouse—especially caterpillars or "worms."

**PROPOXOR (BAYGON®)**—though highly toxic, Baygon is a very effective pesticide against household, lawn, turf, flower, fruit, and garden pests. It is especially effective against resistant cockroaches.

**PYRETHRUM (PYRETHRINS)**—this natural botanical pesticide provides quick knock-down and rapid flushing action of insects. However, pyrethrum does not always kill the insects against which it is used and, thus, is usually mixed with the synergist, piperonyl butoxide, which greatly increases pyrethrum effectiveness.

**RESMETHRIN (SBP-1382®)**—low toxicity plus wide-spectrum effectiveness against many different insects found indoors and outdoors make this a very useful pesticide.

**STIROFOS (GARDONA®; RABON®)**—low toxicity with wide-range effectiveness against many different household, yard, orchard, flowerbed, greenhouse, and garden insect and arachnid pests.

# PESTICIDE
# APPLICATION
# EQUIPMENT

If you plan to practice home pest control on a continuing basis, then you will need to learn about the types of pesticide application equipment available. From this survey, you should be able to select the sprayer(s) and other items that will best serve your own particular purposes. Remember: Correct application of a pesticide is vital if you expect to get effective results. And correct application depends directly upon *you* and the equipment that you use.

The basic types of pesticide application equipment are hand sprayers, mist blowers, aerosol generators/misters/foggers, vaporizers, dusters, and brushes.

### Hand Sprayers

The do-it-yourselfer who plans to practice pest control will find that the acquisition of a hand sprayer of some type is almost essential. Usually you will want the light-weight, portable, basically inexpensive type of hand sprayer that, nevertheless, is powerful and rugged enough to apply pesticides correctly in all situations.

Whether hand-operated or power-driven, sprayers are designed for applying three (3) types of liquid pesticide combination sprays. Space sprays and combination sprays both are directed against visible flying insects such as flies, mosquitoes, gnats, midges, etc. The sprayer and nozzle used in this case must produce *very fine* aerosol-like droplets that will remain suspended in the air for long periods of time for maximum effectiveness. Residual (surface) sprays, however, must be applied as *heavy, coarse* droplets for maximum effectiveness, and thus a different nozzle is required than that used to produce space sprays. Finally, a combination spray should be applied in the form of *intermediate*-size droplets that will remain suspended in the air for a reasonable time, yet eventually settle out to form a fairly good surface film or residue.

**Atomizer, "Flit Gun," or "Fly Sprayer"** This is an old stand-by for the home-treater. It is the smallest, simplest, and least expensive type of hand sprayer in a limited number of cases. These are as follows:

1. Flying insects *indoors*
2. House plant pests
3. Limited area or "spot" treatment for such pests as ants
4. Garden or greenhouse plants that are *small* and *few* in number
5. Small animals

**Portable Compressed Air Atomizer** This type of sprayer also produces a *fine mist* or aerosol spray that is effective for flying insects. This sprayer is designed to operate by compressed air from a hose or portable tank. The compressed air atomizer is limited essentially to the same cases as the hand-operated atomizer.

*Water-Hose Sprayer* This type of sprayer operates by means of a stream of water from a garden hose connected to one end of the sprayer tube. This type of sprayer is *versatile*, allowing you to apply both liquid sprays and wettable powders suspended in water (which requires regular shaking while being applied). This sprayer is especially convenient for treating lawns and turfs, plants, shrubs, hedges, trees, etc. In addition, it can be used to apply grass-killers and other herbicides as well as fertilizers to lawns, turfs, and other areas.

*Hand-Operated (Hydraulic) "Pistol Sprayer"* Squeezing the pistol-plunger handle of this type produces hydraulic pressure which serves to force the liquid pesticide out of the container and through the nozzle. No other attachments are necessary. However, this sprayer is limited to smaller treatment jobs. Additionally, it should be noted that the spray produced by the "pistol sprayer" is *undiluted* by either air or water and thus is much more concentrated. Generally, therefore, this sprayer should not be used to treat animals or most plants.

***Trombone Sprayer*** This type of sprayer also operates by hydraulic force applied by hand through a plunger-and-cylinder which move on each other in trombone fashion. Since the trombone sprayer also produces *undiluted* spray, it too should not be used to treat animals or most plants.

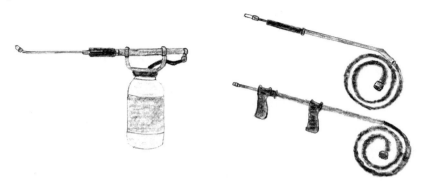

***Garden/Yard Hand-Operated Compression Sprayer*** This type of hand-operated compressed air sprayer is in very common usage throughout the country by professional pest control operators, public health personnel, and do-it-yourselfers. It is one of the most versatile and useful units available, and may be purchased in one of several price categories—depending upon the level of precision and material with which it is constructed. Most expensive are the high-precision stainless steel models designed for professional use. For the average do-it-yourselfer, however, this expensive model is *not* necessary. Rather, the low-to-moderately

priced models are perfectly satisfactory and thus recommended for home use. Some of these models may be constructed partially of plastic, which actually is an advantage since plastic is resistant to the various pesticides and the acids that they produce.

Compressed air sprayers may be fitted with different nozzles, thus increasing the number of methods of treatment. These sprayers may be used both indoors and outdoors. They are especially useful for applying liquid residual insecticide layers against cockroaches and other crawling pests both indoors and outdoors. They also are useful for treating small trees, shrubs, plants, hedges, vegetable gardens, flower beds, greenhouses, and even lawns and turfs.

For continued dependability and effectiveness, compressed air sprayers should be *cleaned after each usage.* Flush the entire unit thoroughly with water, including the hose and nozzle, until clear water appears. After every several usages, the entire sprayer should be disassembled and cleaned thoroughly with a detergent and/or a *solvent* such as: alcohol, benzene, toluene, xylene, gasoline, acetone, or kerosene. At this time, add a few drops of neetsfoot or similar light-weight oil to the leather gasket on the plunger. Also *soak the nozzle* in detergent water or solvent to remove solid pesticide accumulations inside it. NOTE: Do *not* attempt to clean a clogged nozzle with a piece of wire or other metal object. Nozzles are precision machined and can be damaged or ruined when metallic objects are inserted into them.

***Types of Sprayer Nozzles*** Actually, the nozzle is the most important part of a sprayer used to apply liquid pesticides. It is the nozzle alone that determines how the insecticide will be applied—that is, the pattern. Thus, for all practical purposes, the terms *nozzle* and *spray pattern* are synonymous, for each depends directly upon the other.

The four (4) basic types of nozzles and/or spray patterns are illustrated and discussed as follows:

1. *Solid Stream Nozzle*

This type applies a *fine stream* of liquid pesticide. This pattern is recommended to treat cracks and crevices for cockroaches, ants, bedbugs, ticks, silverfish, firebrats, and other hiding insects and arachnids.

### 2. *Flat Fan Nozzle*

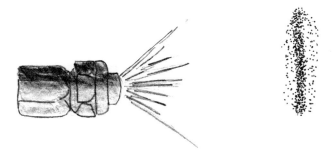

This type sprays insecticide in a *band*. This pattern is recommended for applying *residues* on surfaces and to form *barriers* around yards, dogpens, perimeters, etc.

### 3. *Hollow Cone and Solid Cone*

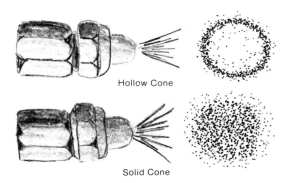

Hollow Cone

Solid Cone

These patterns are recommended to treat *water surfaces* for mosquito and other aquatic larvae; and to treat vegetation, gardens, greenhouses, plants, flowers, etc. for all types of insect and arachnid pests.

### 4. *Adjustable Nozzles*

In some cases it may be possible to purchase a *variable* nozzle which allows application of insecticide in all of the above spray patterns simply by adjusting the nozzle itself. If available, such a nozzle is recommended; otherwise, it will be necessary to purchase the correct type of nozzle for the particular pest problem that you wish to treat.

### Aerosol Generators: Misters/Blowers/Foggers

Aerosol mists or fogs are composed of very tiny *liquid* droplets suspended in air. Because the droplets are extremely small, they have less mass and density and therefore are less subject to the pull of gravity. Thus, they remain suspended for longer periods of time and can be carried great distances by air currents. A very small quantity of liquid pesticide is atomized to form an aerosol. Thus, a maximum of area can be treated with a minimum of insecticide. Aerosols are recommended for all types of flying insects and for tiny pests such as fleas and mites inside homes. Aerosols are particularly good for houses, buildings, orchards, and larger trees, because they provide quick exposure of insecticide to insects over a large area and can be applied quickly. Outdoors, however, aerosols are *less* effective because of air currents.

NOTE: Aerosol fogs or mists are hazardous to your lungs. *Avoid breathing aerosols of any type*, including: insecticides, hair sprays, cosmetics, cleaning agents, and others. Always wear a respiratory or gas mask when applying an aerosol pesticide that requires you to inhale it for more than a few seconds.

*Portable Thermal-Aerosol Foggers* Small, light-weight, portable electric aerosol generators are especially useful for the do-it-yourselfer. Inside, they can be used to fog a single room—or the entire house. Outside, they have a number of uses, such as flowerbeds, gardens, plants, and hedges. These machines produce a large, massive aerosol cloud that rolls over the ground and through the air, thoroughly covering vegetation and plants and killing flies, gnats, mosquitoes, and many other insects. In addition, the barrel of some models can be detached, converting the unit into a sprayer that can be used to apply water-based insecticide sprays. Remember that the heavy, coarse droplet residue layer is recommended for crawling insects as opposed to aerosols for flying insects. Thus, the portable electric fogger with a detachable barrel becomes a very versatile pesticide applicator.

Portable electric foggers are especially useful for fogging or fumigating basements, attics, large rooms or houses, garages, sheds or outbuildings, patios, large-area flower beds, hedges, greenhouses, gardens, and around swimming pools.

***Wheel-Mounted Thermal-Aerosol Misters and Foggers*** For larger and more extensive pest control problems outdoors, a wheel-mounted aerosol fogger operated by a 4-cycle lawn mower engine may be necessary. These are especially useful for treating large areas for flies, mosquitoes, gnats, midges, and pests of orchards and trees. Also, these machines are recommended for fogging large areas of vegetation, flowers, gardens, greenhouses, and hedges.

***Portable Cold-Aerosol Foggers and Misters*** Basically, cold foggers are identical to heat or thermal foggers except for the temperature difference. That is, cold foggers emit a *cold* fog or aerosol cloud that is *less* visible than a thermal fog. Both hot and cold type aerosol foggers are effective, and neither really can be recommended over the other.

***Wheel-Mounted Cold-Aerosol Foggers or Misters*** These units are wheel-mounted and powered by a 4-cycle lawn mower engine. They are very similar to wheel-mounted thermal, or hot, foggers except for the difference in operating temperature. These units are recommended for larger and more extensive pest control applications outdoors.

### Hand Dusters

The hand duster is the third basic type of pesticide applicator. These units are designed to apply *dry dusts* or powders rather than liquids. By their nature, of course, hand dusters are simpler and less costly than sprayers and aerosol foggers. In fact, common household items, such as: perforated cloth, cheesecloth, coffee cans, cocoa cans, salt and pepper

shakers, and plastic squeeze bottles for ketchup and mustard can be converted to pesticide hand dusters. However, if you plan to do your own pest control dusting, you are urged to purchase one of the professional hand dusters illustrated above. These dusters are relatively inexpensive, yet considerably more efficient than the crude hand-made dusters.

*Modified Water Fire Extinguisher Hand Duster* These units are recommended for larger and more extensive insect- or pest-dusting jobs, or for cases requiring the dust to be blown under pressure into hard-to-reach areas. The tank of the modified fire extinguisher is filled to about 80% capacity with insecticidal dust and then pressurized to approximately 100 psi with an air compressor or hose. The delivery hose can be equipped with a narrow spout that allows dust to be blown under high pressure into small holes in walls, floors, etc., and cracks, crevices, voids, subfloors, and beneath cabinets and appliances. These units are especially good for

applying silica aerogels, boric acid, and other dry materials inside walls, subfloors, and other inaccessible areas to control cockroaches, silverfish, and other cryptobiotic (hiding) insects. Additionally, these units may be used to treat smaller trees, shrubs, vegetation, hedges, gardens, greenhouses, orchards, and lawns and turfs.

REMEMBER: An insecticide, regardless of its toxicity, cannot kill insects if it is not applied correctly and thus fails to make maximum physical contact with the pests involved.

# PART II

# PESTS

The biological kingdom, consisting of all living things on earth, is divided into two general kingdoms: animals and plants. Pests are found in both the animal and plant kingdoms. Relatively few, however, are true pests of man; most species are harmless, and many are actually beneficial in the ecological balance of nature that also includes man. Since animal pests are the subject of this handbook, plant pests will not be discussed further.

It is essential to realize that the animal kingdom divides into three (3) fundamental types of living organisms, based upon their respective body plans: (1) VERTEBRATES (with backbones); (2) INVERTEBRATES (without backbones); and (3) ACELLULAR ORGANISMS or PRO-TISTA (consisting of only one cell). To date, zoologists have discovered and named more than 1½ million species (types) of animals, and more species are added to the list each year. The Protista (one-celled organisms) include bacteria ("germs"), protozoa, yeasts, slime molds, and other forms. While many of these organisms are indeed pests of man, they normally are not included in a pest-control book. Therefore, they are omitted from this book also. The pests discussed in this handbook are the larger, more advanced animals found among the VERTEBRATES and INVERTEBRATES.

Zoologists further classify vertebrates and invertebrates into *phylae* (singular: phylum) which can be considered sub-kingdoms. Phylae, or phylums, divide further into sub-phylums. Thus, the phylum Chordata divides into several sub-phylums, one of which is sub-phylum VERTE-BRATA containing the larger and more-advanced animals with backbones, including man.

The vertebrates are one of the two major groups of pests that affect man. Likewise, among the invertebrates, the phylum *Arthropoda* contains the other major groups of pests of mankind. The diagram below illustrates, in very simplified form, how the various groups in the animal kindgom relate to each other and to man.

The two basic groups, vertebrates and invertebrates, are built along entirely different body plans. Thus their external anatomy, internal functions, and life-cycles are very different indeed. Some knowledge of these basic differences between vertebrates and invertebrates is essential to controlling the pests found in each of these two groups.

Most readers already are familiar with the general body plan, internal organs and systems, and the life cycles of the vertebrates—particularly the mammals (class: Mammalia). In mammals, the body plan and organ systems are essentially the same in all species, including man. Even in the lower classes of vertebrates—birds, reptiles, amphibians, etc.—the body functions are quite similar to mammals, with a few minor modifications.

The invertebrate *arthropods*, however, differ greatly from any of the vertebrates. The term "arthropod" means simply, *jointed foot* or *leg*. This group includes *all* jointed-legged animals, which is a vast number indeed. In fact, about 90% of all animals on earth are arthropods! Thus the phylum Arthropoda contains the greatest number and variety of animal species of any phylum in the animal kingdom. And, as you

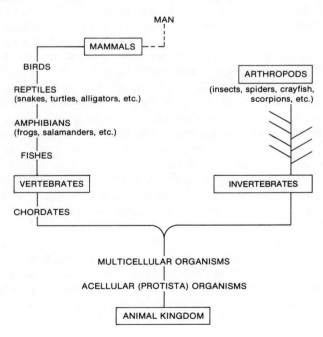

probably suspect by now, the arthropods are the *single largest group of pests* affecting mankind.

Biologists estimate that there are well over *one million* different species (types) of arthropods existing on the earth. Compare this to about 33,000 *combined* species of chordates (vertebrates), including a mere 3,500 species of true mammals, and you will readily appreciate the sheer diversity of the arthropod phylum.

General characteristics of arthropods are as follows:

1. *Segmented* bodies (divided into sections)
2. *Jointed* appendages (legs, antennae, etc.)
3. *Exoskeletons* (skeleton on the *outside*)
4. *Open* circulatory system (no blood vessels)
5. Heart located on the *dorsal* (back) side
6. Nervous system located on the *front* (ventral) side

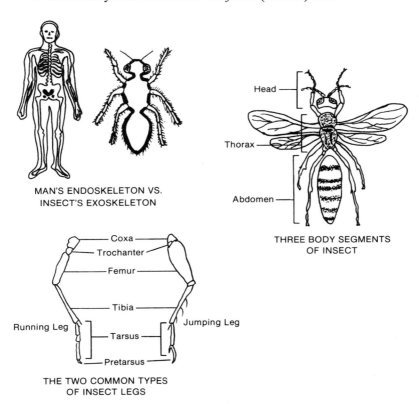

MAN'S ENDOSKELETON VS.
INSECT'S EXOSKELETON

Head

Thorax

Abdomen

THREE BODY SEGMENTS
OF INSECT

Coxa
Trochanter
Femur
Tibia
Running Leg
Jumping Leg
Tarsus
Pretarsus

THE TWO COMMON TYPES
OF INSECT LEGS

The major classes of arthropods are:

1. Centipedes (Class: Chilopoda)
2. Millipedes (Class: Diplopoda)
3. Spiders, ticks, mites, scorpions, etc. (Class: Arachnida)
4. Insects (Class: Hexopoda or Insecta)
5. Lobsters, crabs, shrimp, crayfish, etc. (Class: Crustacea)

**Insects**

The largest single class of animals on earth are the insects. And, as you probably know, insects are the largest single group of pests affecting mankind. The types of insects currently in existence far outnumber those of any other kind of animal. To date, approximately 800,000 species, or types, of insects have been discovered and named. They literally cover the entire earth—from the polar ice caps to the blazing tropics—and are found in the air, on and under the soil, in wood and other materials, and in both fresh and brackish water. Probably insects are the most successful of all land animals. They appeared on earth long before man did, and despite a lapse of some 250 million years, insect fossils are remarkably similar to present day insect species. Thus, insects have survived and flourished, with very little evolutionary change, for hundreds of millions of years.

Most species of insects actually are not pests of man. Only a relatively *few* species are harmful to man and his materials and products. These, however, destroy and contaminate his food, damage his buildings and materials, and transmit serious or deadly diseases throughout the world. Although man has fought insect pests for centuries, he never has eradicated even *one* species from the earth. Man merely tries to control insects, rather than successfully eliminate them.

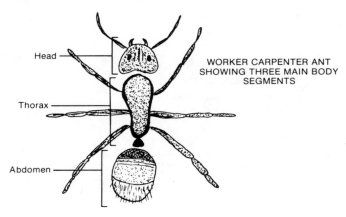

Head

Thorax

Abdomen

WORKER CARPENTER ANT
SHOWING THREE MAIN BODY
SEGMENTS

In addition to the arthropod characteristics given above, most insects have:

1. Three body parts: head, thorax, and abdomen
2. Wings (not all species)
3. Six legs
4. Antennae (not all species)
5. Spiracles (air holes in the body wall for breathing)
6. Simple and/or compound eyes

**Arachnids**

The second largest group of invertebrate pests affecting man are the class of arthropods called *arachnids* (class: Arachnida). Some examples are spiders, ticks, mites, sunspiders, and scorpions. These animals are *not* true insects, but they are similar to insects in size and living habits and are therefore studied along with the latter. Several types of arachnids are serious pests of man.

Although they possess the general arthropod characteristics given earlier, arachnids *differ* from insects in having:

1. Four (4) pairs of legs (8 legs)
2. Two (2) body sections: cephalo-thorax and abdomen
3. Trachea *or* booklungs for breathing
4. Simple eyes
5. No wings
6. No antennae

Brown Recluse Spider

Brown Dog Tick

Scorpion

# ARTHROPODS: ANATOMY AND PHYSIOLOGY

It is important for control purposes that you understand something of the difference between vertebrate and mammalian organ systems and those of insects, arachnids, and other arthropods. These differences are discussed briefly below.

### External Anatomy

*Exoskeleton.* A very basic difference between higher animals (including man) and arthropods is the skeletal system. The human skeleton and that of all mammals is, of course, on the inside. There it serves as the body framework and provides structural support for the muscles, tissues, and organs which surround it. Insects, arachnids, and other arthropods, however, have their skeletons located on the outside of the body where it forms a protective shell called the *exoskeleton*. This four-layered shell consists of a scleroterized (hardened) shell of chitin, wax, and other materials secreted by the body cells beneath it.

This exoskeleton acts as a protective shell, prevents body fluid loss, serves as the framework for the attachment of muscles and tissues beneath it, and even functions as an excretory organ. Thus, in order to exert its toxic effect on the nerves, an insecticide that is not eaten or breathed in by an insect or arachnid must *first* penetrate this armor shell. As you would expect, this requires *time* after exposure of the pest to the chemical.

*Head and Sense Organs.* Insects may have both simple *and* compound eyes. Such a combination gives them exceptional peripheral vision—as much as 360 degrees. This visual perception contributes greatly to their survival. In addition, insects, but *not* arachnids, have antennae which also are sense organs. Still other sensory organs are located on the feet, legs, and other areas of the body. Insects thus are equipped to detect and respond to many forms of stimuli in the environment.

*Types of Mouthparts.* Except for stings, it is in most cases the insect and arachnid *mouth* that directly affects man. Thus, the type of mouth possessed by an insect or arachnid largely determines what kind of damage, or danger, it presents to humans. The basic types of mouthparts found in insects are discussed briefly below.

1. *Chewing*: This is the most primitive and basic type of mouth found in insects. Essentially, this kind of mouth consists of two *mandibles* (jaws) that work laterally or sideways, as opposed to man's jaws which work vertically, or up-and-down. Examples of some insects with chewing mouthparts are cockroaches, grasshoppers, mantids, termites, crickets, beetles, and some forms of lice.

2. *Rasping-Sucking*: Here the mouthparts are modified into an oblong *proboscis* with a tiny file-like rasping surface. This enables the insect to rasp or cut a hole in vegetation and plants, insert the hollow proboscis like a needle, and suck out the juices. Only thrips are known to possess this particular type of mouth.

3. *Piercing-Sucking*: This is one of the basic types of mouthparts found in insects. It also consists of a slender and hollow needle-like proboscis that is used to pierce the outer skin of animals and plants and suck out blood or juices. "Biting" (bloodsucking) insects that attack man and animals possess this type of mouth. Examples are mosquitoes, bedbugs, gnats, some lice, fleas, and others.

4. *Sponging*: Some fly species have this type of mouth. Unable to pierce the skin of plants and animals, these insects must feed on exposed fluids or soft material. Houseflies, for example, regurgitate (vomit) on solid or semi-solid food materials which serves to soften and liquefy it, thus allowing their *sponging* mouthparts to siphon up the liquid or semi-liquid material.

5. *Siphoning*: This type consists of a long slender tube that is coiled up when not in use. Extended, it is used to siphon nectar and juices from plants, flowers, and other vegetation. Moths and butterflies possess this type of mouth.

6. *Chewing-Lapping*: This is a combination-type of mouth that allows some insects, such as bees and wasps, to chew, lap, and

suck up food. Mandibles (jaws) are present and work laterally to chew certain foods. Other parts of the mouth, however, are elongated to form a tongue-like structure that is used to suck or lap up liquids and semi-liquids.

*Thorax.* The mid-part of the insect body is called the *thorax.* In arachnids (spiders, ticks, mites, etc.) the head and thorax are fused together to form one unit, called the cephalo-thorax. The insect thorax sometimes is called the "walking box" or "flying box" because all six legs and the wings are attached to it. The thorax is a very strong and rigid part of the insect exoskeleton that serves the purpose of locomotion. The thorax attaches to the head and abdomen with flexible membranes.

*Legs.* As explained earlier, insect and arachnid legs consist of *jointed* segments. Leg parts are: the *coxa* and *trochanter* which attach to the thorax wall; the *femur* which corresponds to the human thigh; the *tibia* which corresponds to the human lower leg; and the *tarsus* which corresponds to the foot. Leg muscles are located inside the shell-like covering to which they are attached. Insecticidal residues applied to control crawling insects work in most cases by penetrating the insect tarsus as the pest walks across the chemical layer.

*Wings.* Most, but not all, insects possess wings. Usually there are four (4), but true flies (Diptera) have only two wings. Insect wings are simply membraneous outgrowths of the thorax wall, and have tiny veins through which the insect blood (hemolymph) circulates. Exposed wings make an insect highly vulnerable to pesticides.

*Abdomen.* The insect abdomen consists of eleven (11) sections (segments or metameres), although the last one may not be readily visible in all cases. The armor-shell covers the abdomen, but the segments are jointed by thin membranes that allow for movement and flexibility. Each segment has a pair of small air holes (spiracles) located one on each side and leading directly to the trachea (air tubes) which run throughout the body. Generally, the sexual organs are located at the rear of the abdomen.

### Internal Anatomy and Physiology

The approximately one million kinds of insects on the earth vary greatly in size—from very tiny, nearly invisible forms to large specimens measuring several inches. Even the smallest insects, however, possess all of the organs and systems necessary for living—just as a new-born mouse has all of the internal organs possessed by an elephant, except much smaller in size.

***Muscular System.*** All insect muscles are located inside the exoskeleton. Although these muscles are tiny, they form a complex muscular system that may number as many as 2,000 individual muscles in some species. These muscles attach to the exoskeleton in such a way that a very efficient system results. For example, many insects can lift 20 times or more their own weight, while others with specially developed leg muscles can leap or jump inordinately great distances. Fleas, for example, can leap upward a distance of several inches—which, proportionately, compares to a man leaping over a building approximately thirty stories high.

***Nervous System.*** The highly developed insect nervous system coordinates and controls the insect's interaction with the environment. Essentially, the insect nervous system consists of a brain, a nerve cord, and many pairs of enlarged nerve ganglia. The brain is located inside the head with the nerve cord and ganglia lying along the front, or bottom, side of the body.

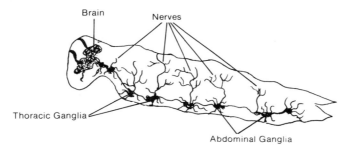

DIAGRAMMATIC DRAWING OF GENERALIZED INSECT NERVOUS SYSTEM

You will recall that most pesticides act as nerve poisons and thus achieve their toxic effects by attacking the well-developed insect and arachnid nervous systems.

***Respiratory System.*** Insects do not have lungs. Rather, *spiracles* (breathing holes) located on the abdominal body wall open into air tubes (trachea) which divide into smaller and smaller tubes called *tracheoles*. These microscopic tubules carry oxygen directly to the tissues. As you can see, this system differs greatly from mammals in which the blood carries oxygen to the tissues. In some cases, insecticidal dusts work by clogging up the breathing tubules of insects as well as the usual way of poisoning the nervous system. Also, fumigant gases and/or aerosol insecticides may enter the insect body directly through the air tubes as well as (more slowly) through the exoskeleton.

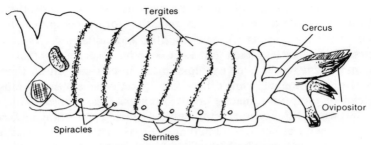

GENERALIZED INSECT ABDOMEN—SHOWING SPIRACLES

RESPIRATORY SYSTEM OF GERMAN COCKROACH
(VENTRAL ARRANGEMENT)

***Excretory System.*** As in higher animals, the insect excretory system discharges waste materials from the body. Insects do *not* have kidneys, however; rather, small tubules, called *malphigian* tubules, serve to remove waste products from the insect blood and discharge these into the digestive tract (gut) for excretion through the rectum and anus. Surprisingly, insects also excrete some of their waste products, particularly uric acid, through the body wall and exoskeleton.

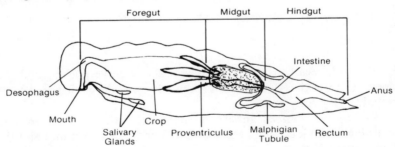

DIAGRAMMATIC DRAWING OF GENERALIZED INSECT DIGESTIVE SYSTEM

***Circulatory System.*** Higher animals possess a *closed* circulatory system consisting of a heart, arteries, capillaries, and veins. Insects and arachnids, however, possess an *open* circulatory system. Essentially, this consists of an open-ended tube that runs the entire length of the back. This tube contracts pulse-like and pumps the blood from the rear of the body to the head, and from there it simply flows back to the rear to be

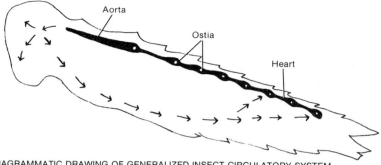

DIAGRAMMATIC DRAWING OF GENERALIZED INSECT CIRCULATORY SYSTEM

pumped forward again. The main functions of the insect circulatory system are: (1) to transport digested nutrients, absorbed through the gut, to the tissues; (2) to carry waste products from the tissues to the malpighian tubules for excretion; and (3) to maintain and/or change the fluid pressure inside the body. Body fluid pressure *changes* are required in insects, particularly during *molting* (to crack the exoskeleton and shed it) and to inflate the wings through which the body fluid or blood circulates.

**Digestive System.** The insect digestive system consists principally of a long tube leading from the mouth through the body to the anus. Attached to this tube (gut) are various glands and organs that aid in food digestion. Like higher animals, insects require a balanced diet of fats, carbohydrates, and protein and a regular supply of water. Some insects, however, can derive their needed water from the food they eat.

The place and type of materials where insects are found can help you determine their identity. Pantry pests, for example, tend to exist in dry cereals, rice, beans, etc. The worm-like larvae of insects found on woolen or hair-like material in the home are likely to be carpet beetles, carpet moths, or perhaps fleas. And cockroaches will of course be found in or near the kitchen, drains, bathrooms, etc. where there is a variety of foods and plenty of water. Crickets inside the house may, in some cases, be found in piles of dirty clothes, soiled socks, or underwear, on which they like to chew.

### Insect Reproduction and Development

All insects develop from eggs. Most develop from eggs that are laid, but a few develop from eggs inside the female body and, thus, are born alive. Because of their armor shell, immature insects cannot grow in the same manner that mammals grow. Instead, insects grow in stages called *molts*, each time the exoskeleton splits open and is shed. A new exo-skeleton quickly forms and keeps the developing insect the same size until the next molt occurs some days or weeks later. Between molts, insects are called *instars*. Most undergo four to eight molts to reach adulthood, but other species may undergo a greater number. Thus, insects change in both size and form as they mature. This process of changing forms is called METAMORPHOSIS.

Basically, immature insects undergo one of four (4) types of change to reach adulthood. These are discussed briefly below.

**Without Metamorphosis.** In this case, growth occurs through a series of molts, with no change in appearance. Each instar appears exactly the same as the others and as the adult, except in size. Silverfish and a few other forms undergo this type of development.

**Gradual Metamorphosis.** Three (3) distinct stages occur during de-velopment: egg, nymph (several molts and instars), and adult. Some examples of insects and arachnids that undergo this type of development are earwigs, grasshoppers, cockroaches, bedbugs, termites, and ticks.

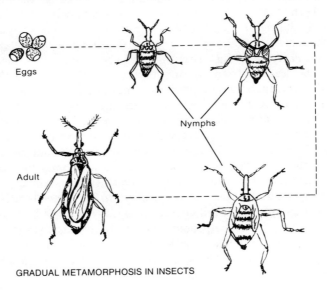

Eggs

Nymphs

Adult

GRADUAL METAMORPHOSIS IN INSECTS

**Incomplete Metamorphosis.** This is an intermediate form of develop-ment between gradual and complete metamorphosis. The developing young are called *naiads* and live in water. The adults, however, live on dry land and are fliers. Some examples of insects that undergo this type of development are mayflies, damselflies, and dragonflies.

**Complete Metamorphosis.** This is the type of insect development that is familiar to most readers: egg, larva (several molts and instars), ɔupa, and adult. Most insects undergo this type of development, in-cluding beetles, wasps, bees, moths, butterflies, flies, fleas, and ants.

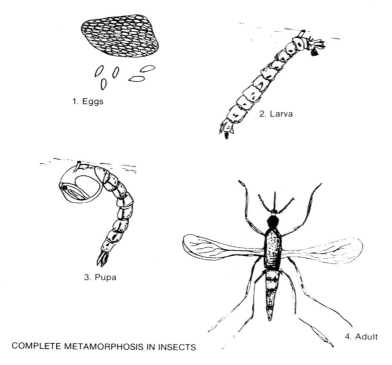

1. Eggs

2. Larva

3. Pupa

4. Adult

COMPLETE METAMORPHOSIS IN INSECTS

# PEST
# MASTER DESCRIPTION
# LIST

The remainder of Part II is devoted to the alphabetized listing and discussion of biological pests. There are more than two hundred of these. Many of the listings that follow are cross-referenced and the reader may need to consult the additional listings for a full understanding of the information given regarding these particular pests.

## ALDERFLIES

*Phylum: Arthropoda    Class: Insecta    Order: Neuroptera*

Alderflies are very similar to dobsonflies and fishflies, except smaller. Usually they are less than one inch long, blackish or gray in color, with a lacewing appearance. Alderflies breed near ponds, streams, and other sources of water. They are attracted to lights at night and, thus, may be pests during the warm months. However, some species of alderflies are important predators of harmful insects, especially aphids, which they help to control. Also, the larvae of some aquatic forms serve as food for fish in fresh water.

### Control

Normally alderflies are *minor* pests only when they swarm around lights at night. Only in cases of severe infestation should you attempt to control them. When necessary, however, control should include (1) treating nearby standing water (pools, ponds, streams, fish containers, barrels, etc.) with diesel fuel, fuel oil, crankcase oil, or the insecticides Abate or Altosid; (2) installing electronic "bug killers" around patio lights and other lighted areas; and (3) applying a *residual* surface insecticide to areas where these insects swarm (2% Sevin, 1–2% Dursban, 2% diazinon, 3% malathion, or 1% Baygon).

## AMERICAN DOG TICKS

*Phylum: Arthropoda    Class: Arachnida    Order: Acarina*

The American dog tick is easily identified by whitish or grayish markings on its back. This tick is very common, and is an important vector (transmitter) of diseases to man. Ticks transmit viruses, rickettsia, bacteria, and protozoa—all of which may cause illness in humans. Ticks also are able to cause paralysis in humans with their neurotoxic (nerve-poisoning) saliva when they remain attached for several days.

Ticks are very hardy and resistant pests, and can go for long periods of time without food or water. With adequate food and water, they may live for years since they have few natural enemies. The tick life-cycle consist of eggs, larvae (6 legs), nymphs, and adults. Most tick species parasitize a wide range of host animals, including man and other mammals, birds, reptiles, and some amphibians.

Ticks have specialized mouthparts which enable them to cut the skin of their hosts. The tick then inserts its barbed hypostome into the wound and sucks blood—for hours or days—until it becomes engorged and swollen, then drops off. The hypostome, which is inserted into the host animal's skin, has *barbs* on its surface that make it difficult to pull the tick off the host.

NOTE: Care must be used in removing ticks from a person or animal; otherwise, the head and mouthparts will be left attached. Normally a tick must remain attached for at least two hours or more before it can transmit to humans any disease organism it is carrying. To remove an attached tick: (1) apply a lit match or cigarette or a hot needle to its body; or (2) apply alcohol, gasoline, acetone, kerosene, or other harsh solvent to its body. As the tick begins to release its grip, very carefully and slowly pull it from the skin. Apply alcohol, mercurochrome, hydrogen peroxide, or other skin antiseptic to the wound.

In most cases a tick bite is not felt by the victim. Once attached, the tick may go unnoticed until discovered by inspection or by accident. When not feeding, ticks generally are found on the ground and in vegetation, where they wait patiently for a victim to come by. Normally ticks are most active during the warm months but a few species, known as "winter ticks," are active even in cold weather.

### Diseases Transmitted by Ticks

*Tick Bite Paralysis.* This condition begins in the legs and extremities, and the flaccidity progresses rapidly through the entire body, producing almost total paralysis of the victim. Eventually, paralysis of the breathing muscles may occur, resulting in death of the victim—if the tick is not found and removed. Tick paralysis results from the neurotoxin found in the female tick's saliva and continually injected into the wound as she feeds. The tick must remain attached for several days to cause paralysis.

Tick paralysis occurs most often in children. A frequent site of attachment is the scalp and back of the neck. Removal of the tick usually produces complete remission of all symptoms within 24 to 72 hours, and the victim recovers with no aftereffects.

*Rocky Mountain Spotted Fever (Tick Typhus).* This is a febrile disease caused by a rickettsial organism transmitted to humans by some ticks, including the American dog tick. Symptoms include: sudden onset of chills, fever, headache, and appearance of blood-shot eyes. In two to four days a rash may appear on the hands and feet and spread over the body, causing spots (blood under the skin)—hence the term "spotted fever." Spotted fever is a serious disease, and may be fatal unless treated with antibiotics. When diagnosed and treated promptly, however, recovery occurs with no aftereffects in nearly all cases.

NOTE: Check your own body and those of your children *daily* during the warm months. If a tick is found attached, remove it carefully, as explained earlier in this section. It is NOT necessary to be concerned nor to call your doctor at this point. First of all, only about 1 to 2% of American dog ticks actually carry the organism that causes spotted fever. Secondly, even if the tick is infected, this does *not* mean that the victim will necessarily become infected, because a number of factors are involved in the infection process. Finally, the organism must incubate within the human body from three to fourteen days before symptoms appear. During this time, nothing can be done. Therefore, note the date on your calendar when the tick was removed from the victim—then watch for the appearance of symptoms over the next three to fourteen days. Should any of the above-listed symptoms appear, contact your doctor at once.

### Control

*On Dogs and Large Animals*—Use the *dust* form of any of the following insecticides: 1% lindane, 0.5–1.0% Co-Ral, 5% malathion, or 5–6%

Sevin; for *liquid washes* use any of the following: 1% Co-Ral; 0.1% lindane, or 1% malathion; for *dips* use any of the following: 0.1% Vapona, 0.5% Dibrom, 1–2% Sevin, 0.2% dioxathion, or 2% ronnel.

**In Homes and Buildings**—Use a *residual* spray: 1% Baygon, 1% diazinon, 1% dioxathion, 1.5% Entex, 0.5% Ficam, 1% lindane, 2% malathion, or 2–3% ronnel. Apply any of these sprays as "spot" treatment to baseboards, window frames, door facings, porches, doorsteps, and other infested areas. 5% Sevin *dust* may be substituted effectively in some cases for the above-listed sprays.

**Yards and Vegetation**—Use a dust, emulsion, or suspension of any of the following insecticides: 1–2% Baygon, diazinon, fenthion, Sevin, Galecron, Fundal, Gardona, Dursban, Ficam, lindane, dioxathion, toxaphene, malathion, or chlorobenzilate.

## ANTS

*Phylum: Arthropoda      Class: Insecta      Order: Hymenoptera*

Ants are highly successful social insects that live in colonies, usually in the ground, and thus are protected from predators and the environment. Ants undergo complete metamorphosis, consisting of egg, larva, pupa, and adult. They live much longer than most other insects. (Workers may live seven years or longer and queens fifteen years or longer.)

Ants eat practically all kinds of food—especially sweets and fatty substances—and thus are serious pests in the home. Additionally, ants bite and sting humans, nest in yards and turfs, steal seeds from seedbeds, eat and chew plants, gnaw holes in fabrics, and even kill young chicks and birds.

### Control

Most ants nest outdoors, but may invade the house to forage for food. But some species may establish their nests inside the house. In any case, long-term control requires that you locate the nest and treat it with insecticide.

**Indoor-Nesting Ants.** These ants may travel some distance from their nest before they appear from a crack or crevice, from between shelves, etc. Use a hand-duster that will force dust deeply into all cracks, crevices, holes, and wall voids. Use 5–10% Sevin. Treat for several feet on each side of the point where the ants appear. Treat baseboards, door moldings, flat surfaces, and cracks along the *entire length* of *any wall* from which ants are seen to emerge. For this treatment, use a *residual* spray of: 1–2% Baygon, 1–2% diazinon, 0.5% Dursban, 0.5% Ficam, or 1–2% pyrethrins.

Also apply this residual band in a 4-inch wide strip along the floor bordering both sides of the treated wall.

As an alternative to dusting or spraying, ant baits may be used. These, however, must be placed out of the reach of children and pets or dispensed from child-proof containers. Recommended ant baits are: Kepone, Mirex, and Baygon.

*Outdoor-Nesting Ants.* These ants may invade the house to forage for food. Follow their trail and attempt to locate the outdoors nest—usually in the ground. Use a liquid garden sprayer to force the chemical into the hole until it is full; or use a dust or wettable powder and wash it into the hole with a garden hose. Finally, spray the entire lower outer wall and foundation, porch, steps, and outer windows of the house with a residual spray. Use any of the following insecticides: 1% Baygon, 0.5% diazinon, 0.5% Dursban, 0.5% Ficam, or 1% malathion. NOTE: When ant nests cannot be located, it may be necessary to spray the entire yard *and* outer portions of the house.

*Carpenter Ants/Ants Nesting under Concrete Slabs.* Techniques for controlling these ants successfully are difficult to apply, and are beyond the capability of most homeowners. Therefore, it is suggested that you employ a professional pest control operator to treat for these pests.

## ANT LIONS

*Phylum: Arthropoda      Class: Insecta      Order: Neuroptera*

Ant lions are large insects that resemble dragonflies, but have longer antennae that are clubbed or knobbed on the ends. Ant lion larvae are called "doodlebugs" and live in pits or holes in sandy, dusty areas. These larvae feed on ants that fall into the pits. Normally ant lions are not serious pests of man, except when they swarm around outdoor lights at night.

### Control

See: *Alderflies* (Control) in this section.

## APHIDS

*Phylum: Arthropoda      Class: Insecta      Order: Homoptera*

Aphids are plant-feeders. They may be serious pests of cultivated plants, causing discoloration, distortion, stunting, wilting, and other similar conditions. Aphids also serve as vectors of plant diseases. Aphids discharge from their rectums a clear, watery substance called "honey-dew" on which ants feed. Ants do, in fact, actually tend aphid "herds" much as a farmer does cows. That is, ants carry aphid eggs to their nests where they remain through the winter, then the ants return the eggs to the plants in spring, where they hatch and develop into a new aphid crop that continues to produce "honey-dew" for the ants. Thus, controlling aphids probably also should involve ant control to some extent.

### Control

Apply to plants attacked by aphids a dust or spray of: 1–2% Sevin, Orthene, malathion, Cygon, Guthion, Dursban, demeton, diazinon, Lannate, Trithion, Di-Syston, Meta-Systox-R, Dimecron, or nicotine sulfate.

### APPLE RED BUGS

*Phylum: Arthropoda     Class: Insecta     Order: Hemiptera*

The apple red bug is a member of the large group of insects known as plant bugs or leaf bugs. Thus, it is closely related to the tarnished plant bug and the four-lined plant bug. The apple red bug is red and black, and about one-fourth of an inch long. It is a serious pest in some apple orchards.

### Control

Use the control methods recommended for *Plant Bugs* in this section.

### ARMY WORMS

*Phylum: Arthropoda     Class: Insecta     Order: Lepidoptera*

Army worms are not true worms at all. Rather, they are the larval stage of a moth. They feed on various grasses and may seriously damage wheat and corn. These larvae commonly migrate in great numbers—hence the term "army" worms.

### Control

Spray or dust surfaces over which army worms crawl with: 1–2% Sevin, 1% Dursban, 2–3% malathion, 1–2% diazinon, 1% Baygon, 2% Bolstar, 2–3% Lannate, Orthene, Ambush, Pounce, Pydrin, or Dimilin.

## ASSASSIN BUGS

*Phylum: Arthropoda     Class: Insecta     Order: Hemiptera*

This group includes a number of "conenose," biting, and blood-sucking bugs. These insects are true bugs and not beetles. They have a narrow, elongated, cone-shaped head with a long beak that is folded beneath the body. This hollow beak is used like a needle to puncture the skin of victims from which the bug sucks blood. One group of these insects is called "kissing bugs" because of their tendency to bite humans about the mouth. Most assassin bugs are predacious on other insects, but others attack and bite humans. These bites can be quite painful. Normally these bugs live outdoors, but the blood-suckers may invade houses and bite humans.

### Control

Outside, spray or dust areas where these bugs occur with: 1–2% Dursban or diazinon, 2–3% Sevin or malathion, 3% DDVP or ronnel, or 2–3% pyrethrins.

## BAGWORMS (Evergreen)

*Phylum: Arthropoda     Class: Insecta     Order: Lepidoptera*

So-called "bagworms" actually are not true worms at all. Rather, they are the larval form of a moth. These moth larvae, or bagworms, construct bags or cases from bits of leaves and twigs in which they pupate on evergreen trees. The wingless female moth lays her eggs in this bag, and may not leave it until the eggs have hatched. Bagworms attack evergreens, particularly cedars, during the summer months and, if not controlled, may cause the trees to turn completely brown or even die.

### Control

Use a pressurized garden/yard sprayer or a pressurized hand duster to thoroughly spray or dust all cedars and other evergreens in your yard. Use a cone nozzle for liquid sprays and apply the nozzle (or duster) *deep* inside the branches and treat the entire tree *thoroughly*. Use any of the following dusts or sprays: 3% malathion or diazinon; 5% Sevin; 2–3%

Orthene, Bolstar, Pydrin, Ambush, or Pounce; 2% toxaphene, Lannate, Baygon, or Dursban; or 3–4% rotenone, methoxychlor, or Dimilin.

An alternate method of control involves pulling the bags off the trees by hand and burning or otherwise disposing of them. This method is tedious and laborious, however, and is recommended only when there are few trees under bagworm attack, or when the infestation is minor.

## BARK LICE

*Phylum: Arthropoda    Class: Insecta    Order: Psocoptera*

These are very small, soft-bodied insects which may, or may not, have wings. They occur in trash, debris, under bark and stones, and on the bark of trees. A few species spin silken webs on the trunks and branches of trees. They feed on dry organic material, molds, and fungi. Some species invade houses where they may cause damage to books by feeding on the starchy material in the bindings.

### Control

Indoors, apply residual sprays, wettable powders, or dusts of: 1% Dursban or diazinon, or 2% methoxychlor or Sevin to surfaces where the insects are. If the lice are actually visible, contact liquid sprays of 1% DDVP, pyrethrins, or SBP-1382® can be used to kill them on the spot.

Outdoors, thoroughly spray (soak) debris, loose soil, bark, and other materials with any of the residual sprays listed above.

## BARNSWALLOW BUGS

*Phylum: Arthropoda    Class: Insecta    Order: Hemiptera*

Barnswallow bugs inhabit bird nests. When these nests are numerous in attics, eaves, or near the house, the bugs may swarm into the house and ferociously bite humans.

### Control

Begin by removing all bird nests from attics, eaves, and trees near or touching the house. For bugs inside the house, treat affected surface areas with: 1–2% malathion, 1% DDVP, 1–2% ronnel or pyrethrins, or 0.5% diazinon. Give special attention to treating attics and eaves.

## BATS

*Phylum: Chordata      Class: Mammalia      Order: Chiroptera*

Bats are mammals—NOT birds. They are fur-covered and give live birth to their young. Bat wings consist of thin tissue stretched between the front and hind legs, enabling the bat to fly. Bats rest during the daylight hours and fly at night, feeding on flying insects which they locate by emitting a high-frequency sound that operates like radar. This enables the bat to accurately locate its flying insect target and nab it in flight. Bats eat large quantities of insects and, thus, are beneficial.

Generally, bats are small—three to five inches long. The female bat usually gives birth to only one offspring, but some may produce two to four young. Although their reproductivity is low, bats generally have a long life span—sometimes twenty years or more.

Bats may lodge in houses, especially in attics, walls, chimneys, or hollow floors. The guano (feces) and urine excreted by a bat colony produce an unpleasant odor. Moreover, ectoparasites (bugs, mites, etc.) from the bats may infest the house and even attack humans. Bats also transmit rabies, which is almost 100% fatal in humans unless treated.

### Control

Generally, bats should be repelled and excluded from a house rather than killed. Killing bats is messy; moreover these mammals actually are quite beneficial in helping to control insects. Bats actively forage for and consume large numbers of insects.

1. *Use a Chemical Repellent.* Naphthalene (mothballs or flakes) or PDB (paradichlorobenzene) flakes will repel bats in most cases. Sprinkle these flakes over the attic or infested area, and pour into cracks in walls and floors. Use three to five pounds of flakes for an average size attic. In semi-open areas, such as porches, hang bags of flakes from the rafters or ceiling to repel bats. NOTE: Naphthalene and PDB volatize (evaporate) quickly. Thus, it may be necessary to apply these repellents more than once.

2. *Bat-Proof the House.* Since bats leave the house approximately one-half hour after dark, wait until the house is clear of bats before bat-proofing it. To bat-proof the house, do the following: (a) cover, or close off, all openings down to one-fourth of an inch in diameter. Wood, oakum, concrete, metal, wire screen, etc. may be used to cover openings EXCEPT VENTS. Here you should use either hardware cloth, cheese-cloth, or fine-mesh wire, with holes no larger than one-fourth of an inch in diameter; (b) leave one or two openings uncovered for several days and

watch these to see whether bats continue to enter and leave the house; (c) when it appears that none do, close off the remaining openings.

NOTE: Bat-proofing should not be done in early and mid-summer because, at this time of year, young bats may be left in the house where they will be likely to die and cause odor.

3. *Special Precautions*: (a) Do NOT handle bats—dead or alive—with your bare hands! Bats commonly transmit rabies which, in humans, is nearly 100% fatal unless treated. Should it be necessary to handle a dead bat, use tongs or gloves; (b) If bitten by a bat, seek medical help at once. If at all possible, capture or kill the bat which inflicted the bite so that your local health department can test it for rabies.

## BEAN LEAF BEETLES

*Phylum: Arthropoda    Class: Insecta    Order: Coleoptera*

Bean leaf beetles belong to the group of ladybird beetles, and are the only members of this group that are destructive to man. The bean leaf beetle, Mexican bean beetle, and squash beetle all are very similar pests of gardens and vegetables. Both the larvae (worms) and adults eat plants and vegetables, and may be very destructive. The bean leaf beetle is yellowish with eight spots on each side of its back. The squash beetle is pale orange-yellow with three spots on the thorax, at least twelve spots arranged in two rows on its back, and a large black dot near the tip.

### Control

Spray or dust affected plants with: 5–10% Sevin; 1–2% malathion, diazinon, Gardona, Guthion, Lannate, or Trithion; or 2% Orthene, Bolstar, Pydrin, Ambush, Pounce, or rotenone.

## BEAN WEEVILS

*Phylum: Arthropoda    Class: Insecta    Order: Coleoptera*

These insects attack both the bean plant in the field and dried beans after they are harvested and packaged. The bean weevil lays its eggs on the pods of bean plants, and the larvae bore into the seeds where they develop into adults. The adult weevils cut tiny holes in the bean seeds and emerge to begin the cycle all over again. This weevil may enter the house in dried beans and once inside, it may breed throughout the year in stored beans or similar food material.

**Control**

Control should begin with the bean plants in your garden or field. Apply to affected plants either the dust or spray form of: 2% rotenone, Gardona, sabadilla, ryania, Sevin, malathion, Orthene, Thiodan, or Lannate.

Indoors, dispose of all contaminated dried beans and other similar foods. Treat all pantry areas, cracks, crevices, shelves, moldings, and other surfaces with a residual spray of: 1.5% Baygon; 1–2% DDVP; 0.5% Dursban, Ficam, or diazinon; 2% malathion; or 3% methoxychlor.

## BEDBUGS

*Phylum: Arthropoda    Class: Insecta    Order: Hemiptera*

Bedbugs are widely distributed and common pests in houses, hotels, motels, and other living quarters where they regularly bite humans and suck blood. Bedbugs are small, flat, oval-shaped, wingless insects, reddish-brown in color and approximately one-fourth of an inch long. Basically, bedbugs are nocturnal (night-feeding) and remain hidden in cracks, crevices, walls, bedsprings, inside mattresses and bedclothes, etc. during the daylight hours. They are not known to transmit disease.

**Control**

Spray bedsprings, bedframes (but NOT bedding), furniture, walls, cracks and crevices with: 1–2% malathion or ronnel; or 1% DDVP or pyrethrins. NOTE: Treat mattresses only along the seams, folds, buttons, and rips and tears. Let mattresses dry completely before using them again.

For areas away from beds, use a pressurized hand sprayer with a fan nozzle for surfaces. For cracks and crevices, use a pin-stream nozzle.

## BEES

*Phylum: Arthropoda    Class: Insecta    Order: Hymenoptera*

Bees are very common insects, and are found almost everywhere, particularly on flowers and other nectar and pollen producing plants. There are at least 3,500 species of bees in North America alone. Bees play an extremely important role in nature—the cross-pollination of plants. The nesting habits of bees are similar to wasps, to which they are closely related. Most bee species are solitary and nest in the ground, in plants, or other similar areas, but two types—honeybees and bumblebees—are highly social and live in colonies ruled by a queen.

Several species of bees often become serious pests of man in and around houses and other buildings. Some persons are highly sensitive (allergic) to bee and wasp venom, and may experience serious illness—or even death—from one or more bee or wasp stings. The major pest-bees are discussed below.

*Honeybees* (see also pp. 22–23). These become pests when a swarm of wild bees enter a house or building through a small void, crack, or hole and establish a nest inside a wall, attic, or other area. Usually honeybees are some shade of yellow, black, or brown with a portion of the abdomen being quite dark. The worker bees are about two thirds of an inch long, and some colonies may have 20,000 to 50,000 bees. The sting of a honeybee can be dangerous to a sensitive (allergic) person.

*Carpenter Bees.* Large carpenter bees are black and yellow and are frequently mistaken for bumblebees, which they closely resemble. Carpenter bees, however, are adept wood borers and bumblebees are not. Carpenter bees may be found boring into the wood of houses, trees, or buildings where they build their nests. Common nesting sites in buildings are eaves, attics, posts, wood siding, wood shingles, doors, windowsills, and other wood surfaces on the outside. These bees also bore into telephone poles, fence posts, and wooden patio furniture. The sting of a carpenter bee can be dangerous to an allergic person.

*Bumblebees.* These are social bees that generally nest underground. They are large, rather clumsy, black and yellow bees which may be confused with carpenter bees. Bumblebees, however, do not bore in wood. Nevertheless, they may become pests when they nest close to sidewalks, paths, buildings, and other areas frequented by humans. The bumblebee sting is venomous, and may be dangerous to allergic persons.

### Control

*Honeybees (see also pp. 22–23)*—First, locate the entrance hole used by the bees to enter the house. Seal off this entrance. If the bees are nesting inside a wall, it is best to use a dust blown into the wall under pressure. It may be necessary to drill one or more holes in the wall in order to apply the dust. (Such holes can be used to treat for cockroaches, crickets, ants, silverfish, and other hiding insects. These holes can be filled in with various caulks or putty and painted over.) The following insecticides are recommended: 5–10% Sevin, Orthene, malathion, Gardona, Furadan, diazinon, or DDVP. Liquid sprays are equally effective, but these may stain the wall. Recommended sprays are: 1% Ficam, 1.5% Baygon, Wasp Freeze, 2% malathion, 1% fenthion, or 1% Dursban. Flying bees can be knocked down with an aerosol or space spray of: DDVP, fenthion, Dibrom, SBP-1382®, Wasp Freeze, Sevin, or malathion.

*Carpenter Bees*—Apply liberal amounts of spray or dust or wettable powder into the entrance holes leading to the carpenter bee nests or galleries, then seal off the entrance hole(s) with putty, caulk, a heavy cloth, or wooden plug. Use any of the insecticides listed for honeybees.

*Bumblebees*—Locate the nest entrance, usually in the ground. Use a pin-stream nozzle to spray a large amount of insecticide into the nest hole or apply a large amount of dust or wettable powder to the nest entrance and wash this into the hole with a garden hose. Now place loose dirt in the nest entrance and pack it tightly with your foot. Use any of the insecticides listed for honeybees and carpenter bees.

## BILLBUGS

*Phylum: Arthropoda*     *Class: Insecta*     *Order: Coleoptera*

Billbugs are large-snout beetles that feed on grasses and foliage. They may become pests in yards, turfs, flower beds, and other areas around the house. The larvae of these insects bore into the stems of plants, and may cause serious damage or loss.

### Control

Spray the yard, turf, flowerbed, or other affected area with: 3% malathion, diazinon, Baygon, Dursban, or lindane.

## BLACKBIRDS

*Phylum: Chordata*     *Class: Aves*     *Order: Passeriformes*

Blackbirds are one of the most pervasive and difficult of all pests to control. These birds regularly swarm in flocks numbering hundreds, even thousands, and present a serious pest and health problem to entire towns, cities, and communities. Besides foraging the countryside and consuming huge quantities of grain, seeds, and other foods, blackbirds are noisy, filthy, and repulsive—particularly in large roosting areas, usually a pine grove or other wooded area.

A blackbird roost that is active for three years or more is considered to present a health hazard to humans in the form of *histoplasmosis*. Humans contact this disease, in most cases, by breathing in the airborne fungal *spores* which are produced in the soil fertilized by the vast amount of droppings in and around a bird roost.

**Control**

Blackbird control normally is a city-wide or community-wide effort, involving many people, health authorities, and even governmental agencies. The individual person is quite limited in his or her ability to control blackbirds effectively. Thus, the suggestions given below are offered only to help alleviate a blackbird problem—not necessarily to solve it.

1. First, thoroughly survey the roosting or nesting area. Determine how big it is and, if possible, the approximate number of birds present.

2. In some cases, particularly if the swarm of birds is not too large, LOUD NOISES produced at timed intervals are effective in scaring the birds away. Some suggested noise devices are firecrackers, guns, carbide cannons, and recorded distress calls played through an amplifier.

3. A weak solution of ammonia water blown into the trees with a mist-blower is sometimes effective in repelling blackbirds. Also, the insecticides endrin, fenthion, and Mesurol have been used to repel and kill birds in some cases.

4. *Poisoning* is probably the most effective bird-control method currently available. Poisoning, however, must be approached with caution and discretion. Killing birds this way may pose legal problems. It also is likely to arouse the furor of avowed bird-lovers. Additionally, *avicides* (bird poisons) may not be available legally in your particular area. In any case, extreme care must be exercised when applying avicides so that children and pets are protected.

Poison baits are recommended. These are: strychnine, Avitrol, DRC-1339® (starlicide), or Ornitrol. Also, in some cases, the insecticides endrin, fenthion, and Mesurol may be effective as baits. Bait foods recommended are wheat, oats, corn, barley, rice, and various seeds.

NOTE: Pre-baiting may be necessary for several days to accustom the birds to the bait. Poison baits should be applied under close supervision and only in uninhabited areas, fields, etc. away from pets and children and non-target birds. Feeding pans or trays containing poisoned bait may be suspended from tree branches in some cases.

## BLACK FLIES/BLACK GNATS/BUFFALO GNATS

*Phylum: Arthropoda*    *Class: Insecta*    *Order: Diptera*

Gnats and flies are members of the Diptera (two wings) order which comprises all true flies. Gnats and black flies are very small black or dark-colored flying insects with a hump-backed appearance. They are vicious biters of man and animals, and the females suck blood. These insects are serious pests in many areas. The larvae of these flies live in streams where they attach to stones and other objects.

### Control

These insects are difficult to control because they breed in running-water streams, often some distance from where they attack man and animals. However, if the breeding area can be located, some degree of control may be achieved by applying liberal amounts of Abate, Altosid, diesel fuel, fuel oil, or crankcase oil to the stream—particularly to rocks, stones, and stationary objects in the water.

For flying adult flies and gnats in and around homes, cabins, tents, and other confined areas, use any of the following aerosol or space sprays: pyrethrins, SBP-1382®, Dibrom, Gardona, DDVP, ronnel, Rabon, malathion, Entex, or thanite.

Recommended repellents are: Rutgers 612, Indalone, or dimethyl phthalate.

## BLACK WIDOW SPIDERS/BROWN WIDOW SPIDERS

*Phylum: Arthropoda      Class: Arachnida      Order: Araneida*

These spiders are recognized easily by their black, brown, or gray abdomen and the red, orange, or yellow hourglass marking on the belly. Immature and male spiders of this group also have additional red markings on the upper side of the abdomen, along with banded legs. Females measure one-third to one-half of an inch in length, with the legs reaching about one and one-half inches. Females are larger than males and after mating the female often eats the male.

The black widow spider lives throughout the Western Hemisphere, and is very common in the southern states from late summer to early fall. The brown widow spider, however, lives mostly in tropical areas and in southern Florida.

Young female widow spiders that have not laid eggs live during the winter in sheltered areas such as buildings, houses, rodent burrows, trash piles, and posts. After mating in spring and summer, the female lays 200–900 eggs in grayish silken balls. The young spiderlings emerge in ten to thirty days. They are cannibalistic, and normally only about one to twelve survive from each egg ball hatch-out. The male spiderlings molt three to six times, and the females molt six to eight times, reaching maturity in two to three months.

The black widow spider bite injects a highly toxic nerve poison. The brown widow spider also inflicts a poisonous bite, but it is far less toxic. The black widow spider is not aggressive and bites only when disturbed.

This bite produces severe symptoms, but death occurs only in about 5% of *untreated* cases. The bite, which feels like a pin-prick, produces excruciating pain within a few minutes. In some victims, the symptoms resemble those of appendicitis: abdominal cramps, rigid or board-like abdomen, heavy sweating, labored breathing, difficulty in talking, and a respiratory grunt.

NOTE: When a person is bitten by a black widow spider, call a doctor or take the victim for medical treatment *at once*. Apply a tourniquet between the bite and the heart, and immerse the bite in ice water. Remember to loosen the tourniquet every five to six minutes.

### Control

Spider infestations usually are evident by masses of webs and egg balls. Use a broom to clean out webs and egg balls. Crush the egg balls and all spiders present.

Both indoors and outdoors, treat spider-infested areas with any of the following insecticides: 0.5% diazinon or Dursban, 0.5% Ficam, 1% Baygon, 1–2% pyrethrins, 0.5–1% resmethrin or SBP-1382®, 3% malathion or Sevin, or 2% DDVP. Use a liquid spray or wettable powder to form a *residue.*

## BLISTER BEETLES

*Phylum: Arthropoda     Class: Insecta     Order: Coleoptera*

These beetles are long and slender with a leathery appearance. Most are black or brown, but a few species are brightly colored. Blister beetles are common and occur on the flowers and foliage of various plants. This beetle secretes the chemical cantharidin, which blisters the skin of man and animals. Adult blister beetles are plant feeders, and some species are serious pests of beets, potatoes, clover, tomatoes, and other plants. These insects may completely defoliate a plant. The larvae are generally beneficial, as they feed on grasshopper eggs. In some cases, however, these larvae parasitize bee eggs within the colony. This occurs when bees, foraging for pollen, accidentally pick up the beetle larvae on their hairy feet and take them back to the colony.

### Control

Apply a dust or residual spray to affected vegetation and other infested areas. Use any of the following insecticides: 2–3% Sevin, malathion, Gardona, Guthion, or Trithion; 1–2% diazinon, Baygon, Dursban, or lindane.

# BLOWFLIES

*Phylum: Arthropoda     Class: Insecta     Order: Diptera*

Blowflies have only one pair of wings and are therefore true flies. This is a large group whose members are very common and abundant. Most species are about the size of the common housefly, but some are larger. Blowflies are metallic blue or green in color. They breed in filth, and the larvae feed on carrion, dung, and similar material. Most maggots found in a dead animal carcass are blowfly larvae.

While these larvae generally feed on dead tissue, a few species, such as the screw-worm fly, attack living animals, penetrating the nostrils, ears, and skin wounds. In the absence of dung and dead tissue, blowflies will breed in fresh and decaying plants and vegetation.

## Control

Control of blowflies should begin with removing all filth and breeding material present such as dog feces, other animal dung, dead birds, rodents, or animals, and decaying plants and vegetation. Normally the adult flies will disappear when all attractive food and breeding materials are removed.

Sometimes, however, it may be necessary to actually treat for blowflies. For larval or maggot control use: 2–3% diazinon, dimethoate, dichlorvos, malathion, fenthion, Cygon, or Gardona in spray form. Dusts also may be used, but results may be slower.

For adult blow flies, use an aerosol or space spray of: DDVP, Entex, dimetilan, Dibrom, ronnel, SBP-1382®, Cygon, pyrethrins, or dimethoate. A residual spray also may be applied to surfaces which these flies frequent. Baits, too, may be used to help control adult blowflies. Use either a dry or liquid bait containing any of the following insecticides: 1–2% dichlorvos, diazinon, dimethoate, trichlorfon, naled, ronnel, or Rabon. Dry baits should be coated with sugar, and liquid baits should be applied with a 10% sugar-water solution (one ounce of sugar dissolved in nine fluid ounces of water).

# BODY LICE

*Phylum: Arthropoda     Class: Insecta     Order: Anoplura*

The body louse is one of three lice types that attack humans (the other two being head lice and pubic or crab lice). These lice all are members of the Anoplura, or sucking lice.

Body lice have been pests of man for centuries, and infestations still occur today throughout the United States and the entire world. When body lice are allowed to produce uncontrolled and infest large numbers of people, they can transmit disease epidemics of typhus, trench fever, and relapsing fever. Like all human lice, body lice also produce a severe skin irritation called pediculosis.

Body lice are very small, flattened, off-white insects without wings. They undergo incomplete metamorphosis: egg, nymph, and adult. Nymphs resemble the adults, but are smaller. When not feeding (sucking blood) the body louse rests on the victim's clothing, and the female normally glues her eggs to fibers of clothing, particularly along the seams and folds. Eggs are rather large and yellowish in color. They may be visible upon very close inspection of clothing, especially with a magnifying glass. Eggs are incubated by human body heat and hatch in about one week. At temperatures above 100 degrees F, the eggs do not hatch. Thus, body lice can be controlled automatically when clothes are worn *intermittently* or stored, because the eggs hatch and the resulting nymphs die from the lack of blood to suck. Body lice thrive *only* in filthy living conditions where clothes are worn daily without laundering and persons do not bathe regularly. The entire body louse life cycle requires about two and one half to three weeks.

NOTE: Most body lice exist on the inner surface of clothing next to the skin. Females congregate along seams, folds, and heavily-stitched areas to lay their eggs. Lice are transferred from person to person by close physical contact.

### Control

Examine the inner surface of clothing, especially underwear, along the seams and folds. Ordinary washing and laundering *will* destroy all stages of body lice on clothes and bedding, except woolens. Woolens should be dry-cleaned, but pressing them along seams and folds also is effective in destroying body lice and eggs.

Normally a body lice infestation is considered to be a medical problem and a physician is engaged. However, body lice can be controlled by washing and laundering of infested clothing, regular bathing, and/or by applying the insecticidal dusts listed as follows: 1% lindane, 1% malathion, 0.2% pyrethrins, or 0.3% allethrin. Apply these dusts on the inner surfaces of underwear and other clothes, and distribute it evenly with your fingers, especially along seams and folds. Also treat the seams of shirts and trousers, as well as socks. These powders have a short residual life and require a second application seven to ten days later.

The following three special shampoos also are recommended for body louse control: A-200 Pyrinate, Cuprex, and Bornate.

## BOLTING CLOTH BEETLES

See: *Cadelles* in this section.

## BOOK LICE

*Phylum: Arthropoda     Class: Insecta     Order: Psocoptera*

These are very small, soft-bodied insects that are strikingly similar to bark lice. Bark lice have wings, however, and book lice do not. Book lice are very active, fast-running insects that often invade houses and other buildings where they may damage books by feeding on the starchy material in the bindings. The immature (nymph) forms of book lice are similar to adults, except smaller.

### Control

Treat infested areas and materials with aerosol or space sprays of: 1–2% methoxychlor, pyrethrins, or malathion; or 0.5% resmethrin. For residual treatment, use: 1% Baygon or diazinon; 0.5% Ficam; 1% Dursban; Whitmire Prescription Treatment #250, 260, or 270; Diazinon-2D Dust, Resipowder Dust, or Ficam Dust.

## BORERS

*Phylum: Arthropoda     Class: Insecta     Order: Coleoptera*

Examples in this group are the clover stem borer, elm borer, locust borer, palm borer, round-headed apple borer, branch and twig borer, flat-headed borer, and pin-headed borer.

These various boring pests are all beetles, the largest order of insects in existence. They attack various trees, fruits, plants, and even lumber. Damage may vary from minor to extensive.

### Control

Treat infested trees and plants with a residual spray. Use any of the following insecticides: 1–2% lindane, Baygon, Dursban, diazinon, or malathion. Cut lumber, such as flooring, fencing, etc. should be treated with a wood preservative, such as: PCP, creosol, or Vikane.

## BOXELDER BUGS

*Phylum: Arthropoda     Class: Insecta     Order: Hemiptera*

The boxelder bug is about one half of an inch long, blackish or dark in color, with red markings. It feeds on boxelder and other similar trees. This bug may invade houses in great numbers, especially in the fall.

### Control

Outdoors, treat trees and ground areas with a spray or dust of: 1% diazinon, Baygon, or Dursban; 0.5% Ficam; or 2% malathion or Sevin.

Indoors, use: 0.5% DDVP or pyrethrins; or 0.3% SBP-1382®. Use an aerosol or space spray to fill the entire room, then close it off for several hours. Wall voids and other such areas should be treated with a 5–10% malathion or Sevin dust.

## BREAD BEETLES

See: *Cadelles* in this section.

## BRISTLETAILS (Firebrats)

*Phylum: Arthropoda     Class: Insecta     Order: Thysanura*

Bristletails are very small, wingless insects with two or three tail-like extensions and very long antennae. They are fish-shaped and silvery, brownish, or dark in color. They commonly live in houses, buildings, and restaurants where they feed on starchy materials, such as wallpaper paste, cloth, bookbindings, and stored food. Look for these insects in dusty attics, particularly in boxes of old books and papers, in bookcases, and occasionally in bathtubs where they get trapped looking for water.

### Control

Use any of the following aerosol or space sprays in areas of infestation: 1% Baygon, DDVP, diazinon, or Dursban; 0.5% Ficam; 2% pyrethrins, malathion, Sevin, or methoxychlor. For wall voids and other hard-to-reach areas, use silica aerogels or dust forms of the above-listed chemicals.

## BROWN DOG TICKS

*Phylum: Arthropoda    Class: Arachnida    Order: Acarina*

Ticks are not true insects; they are arachnids. The brown dog tick is a very common and abundant pest in most areas of the United States. It is a reddish-brown tick that attacks dogs and other mammals, but rarely bites man. The brown dog tick is not known to transmit any disease in the United States.

This tick is very common in and around dog pens, kennels, and houses. It often swarms in great numbers, invading porches and outside areas of the house. This tick is very difficult to control.

### Control

Effective control usually requires treatment of all dogs and the surrounding yard, grounds, kennels, buildings, and even the human living quarters. For treating dogs, the systemic insecticide ronnel is recommended. Dog and cat collars containing DDVP also are recommended. For yards, grounds, and outbuildings, use any of the following insecticides: 2% dioxathion (Delnav), Baygon, toxaphene, Dursban, Ficam, or lindane; 3–5% Sevin; 2–3% chlordimeform, diazinon, malathion, or fenthion.

Inside, use: 0.25% Ficam; 0.5% lindane; 2% malathion or ronnel; or 1% Baygon. Use these sprays as spot treatment to baseboards, floor and wall cracks and crevices, window frames, door facings, porches, steps, and other areas infested by ticks. Residual sprays and dusts are recommended in all cases.

## BROWN RECLUSE SPIDERS

*Phylum: Arthropoda    Class: Arachnida    Order: Araneida*

The brown recluse spider is one of three poisonous spiders in the United States (the other two being the black widow and brown widow spiders). The brown recluse spider is common in houses, particularly in closets where it hides in folded and hanging clothes. It also may hide in bedding, wall voids, attics, and other dark and reclusive areas—hence the name brown "recluse."

The body-length of this spider is about one-half of an inch. It is light brown to dark brown, with a fiddle-shaped marking on the back (cephalothorax). This spider is very shy and does not bite unless provoked. Eggs are laid in off-white silken cases that are about one-third of an inch in diameter. In summer, the spiderlings emerge in twenty-four to twenty-six days. They develop slowly, maturing in seven to eight months. The adults may live five years or longer.

NOTE: Both the male and female brown recluse spiders inflict poisonous bites. Reaction-time to the bite varies from immediately to as long as eight hours. Typically, a small blister forms, and a large area around the bite becomes swollen and sensitive. The victim becomes restless, feverish, and has difficulty sleeping. Tissue around the bite dies and gradually sloughs off, sometimes leaving a pit or hole in the flesh. Six to eight weeks or longer may be required for healing to occur.

ALWAYS consult a doctor when anyone is bitten by any kind of spider, because allergic reaction to spider bites varies from mild to serious.

### Control

When spiders are visible, a space or aerosol spray is recommended. Use any of the following insecticides: 1–2% DDVP, 0.5% pyrethrins or SBP-1382®. When spiders are not visible, use a residual insecticide of: 1% Baygon, diazinon, or Dursban; 3% malathion or Sevin; or 2% ronnel. The dust form of any of these chemicals can be used in crawl spaces, attics, shelves, wall voids, and other out of the way areas.

Outdoors, use the same insecticides listed above.

## BROWN WIDOW SPIDERS/BLACK WIDOW SPIDERS

See: *Black Widow Spiders/Brown Widow Spiders* in this section.

## BUFFALO BUGS (Carpet Beetles)

*Phylum: Arthropoda*     *Class: Insecta*     *Order: Coleoptera*

These are tiny, oval beetles, blackish in color with minute white markings and an orange band along the back. They are common insects outdoors, on plants and trees, and indoors where they infest carpets, textiles, and woolens. Eggs hatch in ten to eighteen days, producing reddish-brown larvae with a "hairy" appearance. The larvae appear to run rather than crawl. They molt about six times, and the pupae may occur in cracks and crevices beneath the carpet. Upon emerging from pupation, the adult carpet beetles lie quiescent for some eighteen days before becoming active.

Adult carpet beetles DO NOT feed on woolens and similar materials. Rather, it is the larvae that feed on these materials and cause damage. Moreover, the adults move about and may appear in areas away from where the larvae are feeding and causing damage. The damage-causing larvae feed in dark, secluded places such as closets, furs, woolens, bits of hairy material, and carpets. They also live in lint, behind door facings and baseboards, in some upholstered furniture, in air ducts, and some kitchen cereals.

### Control

Use a knife blade, spatula, or similar tool and a flashlight to search for the beetle larvae. Remove bird and wasp nests from attics, eaves, and such areas, as well as masses of animal hair from corners, vents, beneath furniture, and other such areas. Remove all infested material except whole carpets. Use one of the following space sprays for spot treatment only: 2% malathion, 5–7% Perthane, 1% lindane, or 0.25% Ficam. Treat baseboards, beneath furniture, and in the seams, folds, buttons, and cracks of furniture. A water-based spray is recommended to avoid possible damage to carpets and fabric materials.

For non-fabric areas, use any of the following insecticides: 0.5% Dursban or diazinon; 3% malathion; 2–3% DDVP; 1% Baygon or diazinon dust.

## BUMBLEBEES

See: *Bees* in this section.

## CABBAGE WORMS

*Phylum: Arthropoda*    *Class: Insecta*    *Order: Lepidoptera*

The cabbage worm is actually the larval form of a butterfly. These are medium-sized white butterflies with black markings. The larvae, or worms, do considerable damage to cabbage and other related vegetables and plants.

### Control

Treat affected plants with a spray or dust. The following insecticides are recommended: 2–3% malathion, Sevin, Gardona, Orthene, Pounce, Ambush, Pydrin, Lannate, or rotenone.

## CADELLES
### (Bread Beetles or Bolting Cloth Beetles)

*Phylum: Arthropoda*     *Class: Insecta*     *Order: Coleoptera*

The adult cadelle is about one-third of an inch long and shiny black. The larvae are about five-eighths of an inch long, dirty white in color, with dark brown heads. This pest occurs commonly in rice and flour mills, but home infestation may occur in limited areas, such as a single kitchen cabinet or even one box of stored food. This beetle also is found in ground cereals, packaged foods, corn, oats, nuts, spices, and fruits. Both adults and larvae eat holes in sack containers and even in wood.

### Control

Begin by locating the infested area—perhaps a single kitchen cabinet or a single box of stored food. Check cracks and crevices, the insides of containers or foods such as cereals, beans, flour, peas, spices, dried fruits, and similar material. Destroy the infested material and treat shelves, drawers, and cracks and crevices with a residual spray of: 1% Baygon, 0.25% Ficam, 0.5% diazinon or Dursban, 2% malathion, 2–3% methoxychlor, 2% Sevin, or 1–2% DDVP. Allow the wet spray to dry completely, then cover the sprayed area with fresh shelf paper.

## CADDIS FLIES

*Phylum: Arthropoda*     *Class: Insecta*     *Order: Trichoptera*

Caddis flies are slender, elongated, moth-like insects, varying from one-tenth of an inch to about one inch in length. They are dull-colored, and move in a jerky, erratic flight pattern. The adults rest during daylight hours, and fly about at night. They are strongly attracted to lights and, thus, become pests in and around patios, pools, porches, and other areas during the warm months.

### Control

Caddis flies breed in water. If the source can be located, the immature stages can be treated with any of the following chemicals: Abate, Altosid, diesel fuel, fuel oil, or crankcase oil. Flying adults can be treated with an aerosol or space spray of: 2–3% Baytex, Entex, dimetilan, DDVP, Dibrom, malathion, diazinon, SBP-1382®, Lethane, Cygon, ronnel, or Gardona.

Surface areas around lights may be treated with a residue of any insecticide listed above. Also, electronic bug killers may be installed around lights to help control these flying insects.

# CANKERWORMS

*Phylum: Arthropoda      Class: Insecta      Order: Lepidoptera*

Cankerworms are the larvae of a moth group which also includes the measuring worm or inch worm. Cankerworms feed on various deciduous trees, and may cause serious defoliation. The spring cankerworm and fall cankerworm are two common species of this group.

### Control

When many trees, or large trees, are involved, it is difficult to treat effectively for cankerworms, since treatment requires thorough spraying of the leaves and branches. Usually a tractor-powered or portable engine-driven fogger or mister is required for effective treatment of trees. The following insecticides are recommended as contact, space, or residual sprays: 2–3% malathion, Sevin, diazinon, Gardona, Guthion, Dimilin, Orthene, Pydrin, Ambush, or Pounce; or 1–2% Baygon or Dursban.

# CARPENTER BEES

See: *Bees* in this section.

# CARPET BEETLES

See: *Buffalo Bugs* in this section.

# CARPET MOTHS (Tapestry Moths)

*Phylum: Arthropoda      Class: Insecta      Order: Lepidoptera*

The carpet, or tapestry, moth is rare in the United States. Nevertheless, this pest can cause substantial damage when there is a severe infestation. The carpet moth wingspan is about one and three-fourths inches. The basal one-third of the forewings is black and the outer two-thirds creamy white. The hindwings are uniformly gray. The larvae of this moth prefer coarse, heavy fabrics, and construct burrows throughout carpets, tapestries, felt, furs, skins, old woolens, feathers, etc. The carpet moth may also attack and damage wallpaper.

**Control**

See: *Buffalo Bugs* (Control) and *Casemaking Moths* (Control) in this section.

## CASEMAKING MOTHS

*Phylum: Arthropoda*     *Class: Insecta*     *Order: Lepidoptera*

These are small, somewhat brownish, plain-colored moths with three small, dark spots on each front wing. They are also called clothes moths by some people. Adults have a wing-spread of less than one-half inch. They rarely fly, and live in dark corners, closets, and within the folds of clothes and fabrics. They tend to nest in clothing, carpets, rugs, upholstery, fabrics, piano felts, brush bristles, blankets, pet hair masses, furs, lint, and woolens. The larvae are casemakers, and feed on fabrics, woolens, and similar materials which they may damage.

### Control

Household cleanliness is necessary to control clothes moths and/or casemaking moths. This includes thorough and frequent housecleaning, vacuuming, brushing, airing of clothes and woolens, and sometimes dry cleaning. Fabric pieces, carpet pieces, feather pillows, and similar materials that are infested should be discarded, and areas around baseboards, behind doors, under radiators, and inside furnaces and vents should be thoroughly vacuumed.

Moth repellents—PDB or naphthalene (mothballs)—should be applied routinely. Mothproofing chemicals also can be applied as spot treatments, but only to heavily-infested areas. The following chemicals are recommended: 1–2% malathion, 4–5% Perthane, or 0.5% lindane. For carpets, apply the spray around baseboards, and under furniture. For upholstered furniture, apply spray around buttons, seams, cracks, and folds. A water-based spray is recommended, applied with a fan or cone nozzle. NOTE: Do NOT soak fabrics. Instead, apply the spray lightly and rapidly. Keep children and pets away from sprayed materials until they are completely dry.

For visible adult moths, use any of the following *aerosols* or *space* sprays: 1–2% DDVP or SBP-1382®; 2% pyrethrins; 2–3% methoxychlor; 0.5% Ficam; or 2% malathion.

Non-fabric areas may be sprayed with: 0.5% Dursban, 1% diazinon, or 2% malathion. For inaccessible areas such as wall voids and subfloors, any of the following dusts may be used: 2% diazinon; 5% malathion, Sevin, or methoxychlor; 1–2% Ficam; or silica aerogel.

# CATERPILLARS

*Phylum: Arthropoda     Class: Insecta     Order: Lepidoptera*

Caterpillars, called worms, are the larvae of various butterflies and moths. Caterpillars are very common, and exist on various plants, trees, and vegetation. Some pupate in cocoons, while others do not. Most caterpillars feed on the external foliage of plants and trees, but a few live inside leaves; some form galls, and others bore into the fruit, stems, and parts of plants and trees. Some of the most common and destructive caterpillars are discussed below.

*Hornworms.* These caterpillars have a soft spine-like protrusion near the rear end of the body. Some hornworms attack tomatoes, tobacco, and other plants. They may leave the plants and pupate in the ground.

*Oak Moth Caterpillars.* One species of oak moth caterpillar occurs in California. These caterpillars feed on oaks and occasionally on other trees to which they may cause considerable damage.

*Measuring Worms or Inchworms.* These caterpillars are easily recognized by their looping, measuring movement. They feed on many different plants and trees, and some species seriously damage orchard and shade trees.

*Tent Caterpillars.* Eggs of this moth lie dormant during the winter and hatch early in the spring. These larvae, or caterpillars, are active by late May or early June. They build and live in silken tents in the forks of tree branches. These caterpillars feed on apple, cherry, and related trees, which they defoliate completely.

*Woollybears.* These caterpillars are very hairy in appearance—hence the name "woollybears." A few species feed on trees and shrubs.

*Saddleback Caterpillars.* These larvae have short legs, and move about rather like slugs. A dark spot on the back resembles a saddle. Some species possess stinging hairs which can prove painful if handled. These caterpillars feed on various trees and shrubs.

## Control

Treat infested plants, vegetables, shrubs, and trees with a dust, wettable powder, emulsion, or spray. The following insecticides are recommended for most caterpillars: 2–3% Orthene, Bolstar, Pydrin, Ambush, Pounce, Sevin, malathion, toxaphene, rotenone, or Lannate; or 1–2% Baygon, Dursban, diazinon, or Dimilin.

## CATTLE GRUBS

*Phylum: Arthropoda*     *Class: Insecta*     *Order: Diptera*

Cattle grubs are the larvae of heel flies or ox warble flies. While these larvae normally occur on cattle or deer, they also may parasitize horses and humans. Adult flies lay their eggs on the hair of cattle. The eggs hatch in approximately one week, and the tiny larvae (grubs) crawl deep into the hair and bore directly into the skin. The grubs migrate through the body, lodging finally just under the skin of the back, where they cut holes through which to breathe. Here they grow and develop. Finally they burst through the skin, fall to the ground, crawl away, and pupate. In four to five weeks, the adult flies emerge, and the cycle is repeated.

### Control

Add the systemic form of ronnel to feed or to salt blocks. Apply directly to the hair and skin of affected animals any of the following insecticides: coumaphos, Ruelene, trichlorfon, Warbex, or Ciodrin.

## CATTLE LICE

*Phylum: Arthropoda*     *Class: Insecta*     *Order: Coleoptera*

These are blood-sucking ectoparasites that belong to the general group of sucking lice known as the Anoplura. Five (5) species of these lice are known to attack cattle. Other species attack hogs, horses, mules, donkeys, dogs, sheep, and domestic rabbits.

### Control

Apply to affected animals any of the following insecticides: coumaphos, Ruelene, trichlorfon, Warbex, or Ciodrin.

## CATTLE TICKS

*Phylum: Arthropoda*     *Class: Arachnida*     *Order: Acarina*

Several species of ticks parasitize cattle and other animals, from which they suck blood.

**Control**

Apply to the hair and skin of affected animals a dust containing: 1% lindane, 0.5–1% Co-Ral, or 3–5% malathin or Sevin. The following liquid washes also can be used: 1% Co-Ral, 0.05% lindane, 1% malathion, or 1–2% Sevin.

# CENTIPEDES

*Phylum: Arthropoda       Class: Chilopoda       Orders: Scutigeromorpha,*
*Lithobiomorpha, Scolupendromorpha, and Geophilomorpha*

Centipedes are slender, elongated, worm-like animals with fifteen pairs or more of legs, one pair per body segment. Centipedes are common animals, found in soil, debris, under bark, in rotten wood, and similar dark, damp places. They are active, fast-moving animals that eat small insects and spiders. Occasionally centipedes invade houses where they become pests. Smaller ones are harmless, but larger centipedes can inflict a very painful bite. The bite is non-poisonous, but a local allergic reaction may occur. Centipedes do not damage food or household furnishings.

**Control**

Inside, apply a residual spray to flat surfaces, cracks, crevices, base-boards, and to dark, damp areas. The following insecticides are recommended: 2–3% malathion or Sevin; 1% Baygon; or 0.5% Dursban, diazinon, or Ficam. Inside wall voids, subfloors, and other inaccessible areas, a pressurized hand-duster may be used to apply a residual dust of 10% Sevin, 3% malathion or diazinon, or 1% Ficam.

Outside, use any of the above-listed residual sprays to thoroughly treat soil, yards, trash and debris piles, house and building foundations all around, steps, porches, and other areas where centipedes are seen.

# CHEESE MITES

*Phylum: Arthropoda       Class: Arachnida       Order: Acarina*

This mite infests cheese, flour, grain, and other types of stored food. When infested products are handled by humans, this mite can cause a mild skin irritation called dermatitis.

**Control**

Discard infested food products. Remove dishes, shelf paper, boxes, etc. from affected cabinets, shelves, cupboards, storage rooms, and other areas. Use any of the following insecticides in contact or residual spray, or in dust form: 2% malathion, methoxychlor, or Sevin; 1% Baygon; 0.5% Dursban or diazinon; 0.25% Ficam; or 1–2% DDVP or pyrethrins.

## CHEESE SKIPPERS

*Phylum: Arthropoda     Class: Insecta     Order: Diptera*

Cheese skippers are the larvae of a skipper fly. The adult flies are metallic bluish to black in color, and are about one fourth of an inch long. The larvae, or cheese skippers, are serious pests of cheese and preserved meats. These larvae move by jumping or skipping—hence the term "skipper."

**Control**

Control should begin with the adult flies which produce the larvae, or skippers. Fly-proof openings to storage rooms and other areas where meat and cheese are kept with 30-mesh screen wire or smaller. Thoroughly clean scraps, crumbs, bits of cheese and meat, and grease from all storage areas, shelves, tables, floors, bins, etc. In some cases, it may be helpful to thoroughly rinse all stored cheese and meat with clear water to remove eggs and larvae.

When adult flies are present, use an aerosol or space spray of: pyrethrins, DDVP, SBP-1382®, malathion, Gardona, Dibrom, or ronnel. Fly baits also may be used: dichlorvos, malathion, Cygon, trichlorfon, or Dipterex.

## CHICKEN FLEAS

*Phylum: Arthropoda     Class: Insecta     Order: Siphonaptera*

Chicken fleas are distributed widely, and sometimes attack humans. This flea commonly lives on fowl, especially chickens, and may infest bird nests. When bird nests are located in, on, or near the house, chicken fleas may invade the interior and bite the human inhabitants.

**Control**

Indoors, check attics, eaves, wall voids, and other areas for bird nests and remove them. If fleas are swarming inside the house, vacuum the floors, furniture, and carpets *daily* and discard the vacuum bag each time. Treat attics and eaves with a residual spray of: 0.5% Ficam, 1% ronnel, Whitmire Prescription Treatment #585, 2% malathion, 0.5% Dursban or diazinon, or 1.1% Baygon. Finally, close the house tightly, remove all human inhabitants and pets, and release two or three aerosol insecticide bombs. Allow this aerosol treatment to work for three to four hours before re-entering and venting the house.

If fleas still persist, treat the carpets, floors, furniture, baseboards, etc. with water-based 0.25% Ficam.

## CHICKEN MITES

*Phylum: Arthropoda*    *Class: Insecta*    *Order: Acarina*

This mite parasitizes chickens, domesticated fowl, and wild birds. Often it invades homes from chicken houses nearby, or from bird nests located in or near the house, and bites humans. Nests of pigeons, starlings, and sparrows are likely to produce chicken mites that may invade houses and bite humans.

**Control**

Remove all bird nests from attics, eaves, and other areas in or near the house. Apply thoroughly a residual spray to the entire attic, eaves, and other affected areas. Also treat chicken houses and roosts with this spray. If mites have invaded the home, close the house tightly, remove all inhabitants and pets, and release two or three aerosol insecticide bombs. Allow the bombs to work for three or four hours before re-occupying and venting the house. Recommended chemicals for sprays and bombs are: 1–2% malathion, 1% diazinon or Baygon, 0.5% Dursban, 1–2% pyrethrins or resmethrin, or 2% Sevin.

Outside areas may be treated with any of the following insecticides or miticides: Acaraben, Kelthane, Mitox, Thiodan, Omite, Ovex, Aramite, Plictran, Pentac, Galecron, or Fundal.

## CHIGGERS

*Phylum: Arthropoda*    *Class: Arachnida*    *Order: Acarina*

Chiggers are the larvae of harvest mites. These mites lay their eggs on the ground, and chiggers hatch and move about until they find a host from which to suck blood. Contrary to popular belief, chiggers do *not* bury into the skin; rather, they puncture the skin and feed through a small tube, releasing saliva into the wound, which causes severe irritation, itching, and welts. Chiggers are extremely small mites, and may be invisible to the naked eye. They are very common pests throughout the United States, and are active year-round in the southern states. They live in grass, weeds, and vegetation.

### Control

Use an emulsion, contact, or residual spray to thoroughly treat lawns, grass, weeds, and vegetation. The following insecticides are recommended: 1% diazinon; 2% Sevin or malathion; 0.5% Dursban or Baygon; or 1% lindane or toxaphene.

Repellents specified for chiggers also may be used.

### CHINCH BUGS

*Phylum: Arthropoda     Class: Insecta     Order: Hemiptera*

The chinch bug is a member of a large group of seed bugs. It possesses a long needle-like beak, usually folded beneath the body. The chinch bug is less than one-fourth of an inch long, black, with white front wings. These bugs spend the winter as adults in grass, leaves, fence rows, and other concealed areas, and emerge in April to begin feeding on small grain.

The chinch bug is the most injurious member of this large seed bug group. Chinch bugs readily attack wheat, corn, and cereals. Often they swarm in turf grasses and leaves, and may invade houses in great numbers.

### Control

Outside, thoroughly treat the lawn, grounds, and grasses with a residual spray or dust of: 1–2% Dursban, Baygon, diazinon, or Ficam; 2–3% malathion or Sevin; or 1–2% Aspon. Also treat outer walls of the house, window sills, doorsteps, porches, crawl spaces, door facings, and other affected areas.

Inside, use a contact or residual spray of: 1.1% Baygon, 2% Baygon bait, 0.5% Dursban, 2% malathion, 3% Sevin, or 0.25% Ficam. Additionally, or alternatively, aerosol insecticidal bombs may be used after closing the house tightly and removing all inhabitants and pets. Aerosol bombs vary in strength, insecticidal contents, and effectiveness, however, and you must select one specified for "chinch bugs" or "bugs."

# CHIPMUNKS

*Phylum: Chordata     Class: Mammalia     Order: Rodentia*

Chipmunks are small squirrels that may be confused with true ground squirrels, which are larger. Chipmunks normally are harmless, colorful little animals that enhance outdoor enjoyment. However, at times, they burrow under buildings, in flower beds, lawns, turfs, golf courses, and other such areas, thus making themselves pests. Chipmunks live in underground burrows, inside building walls, in subfloors, and in woodpiles. They are active from March until October, and feed on seeds, nuts, grain, fruit, and insects.

### Control

First, try repelling these animals by injecting liberal amounts of PDB or mothballs into their burrows, into walls and floors where they nest, in woodpiles, and other such areas. Ground burrow entrances should be packed with dirt after the repellent is injected.

Trapping also is effective against chipmunks. Use rat snap traps, small bait traps, small-animal wire traps, cage traps, or Hav-a-Heart traps. Place these traps near rock piles, burrows, logs, woodpiles, and other infested areas. NOTE: Use gloves or tongs to handle both dead and live chipmunks, since they may transmit disease.

# CICADAS

*Phylum: Arthropoda     Class: Insecta     Order: Homoptera*

Cicadas, also known as "locusts," are large blackish, greenish insects, one to two inches long. They are very common, appearing yearly in July and August, but are more likely to be heard than seen because they are arboreal (tree dwellers). Several cicada species occur in the United States. Periodical cicadas occur mainly in the east, have a life cycle of thirteen to seventeen years, with adults present only in certain years. Seventeen-year cicadas are mainly northern in distribution, and thirteen-year cicadas occur primarily in the south.

Female cicadas lay their eggs in the twigs of bushes, trees, shrubs, etc. The injured twigs, in turn, usually die and drop off, allowing the eggs to hatch into nymphs on the ground. Nymphs are stout-bodied, brownish insects with *wide* front tibia (legs) that climb the trees to molt. Large numbers of egg-laying cicadas may seriously damage young trees.

**Control**

Apply to affected trees and shrubs a residual or contact spray, or dust, of: 2–3% Sevin or malathion; 1% diazinon or Baygon; or 0.5% Dursban.

## CICADA KILLERS

*Phylum: Arthropoda*     *Class: Insecta*     *Order: Hymenoptera*

The cicada killer is a large black wasp with yellow stripes circling the abdomen. The larger cicada killers may be one and one-half to one and three-fourths inches long. These wasps prey on cicadas, which they kill or paralyze with their sting, and thus provide their nests with food. Cicada killers often are mistaken for hornets. However, cicada killers dig and nest in the ground while the smaller hornets build large visible nests above ground. Like most wasps, the cicada killer packs a vicious sting that can be dangerous to allergic persons.

**Control**

Locate the entrance hole(s) to the underground nest(s). Spray into the hole(s) a large amount of liquid insecticide, or wash in with a hose a large amount of insecticidal dust or wettable powder, then pack dirt tightly into the entrance. When burrowing is extensive, treat the entire lawn with a residual spray. The following insecticides are recommended: 1–2% dichlorvos, Baygon, diazinon, Dursban, Ficam, Baytex, Sevin, Dibrom, dimetilan, Lethane, or SBP-1382®.

## CIGARETTE BEETLES

*Phylum: Arthropoda*     *Class: Insecta*     *Order: Coleoptera*

These are cylindrical to oval, pubescent, light-brown beetles that fly readily. Less than one-half of an inch long, cigarette beetles are common and destructive pests of dried tobacco, museum specimens, insect collections, and stored foods. They also infest books, cottonseed meal, rice, dried fish, drugs, seeds, dried plants, paprika, pepper, ginger, and flax.

**Control**

Indoors, discard all infested material and foods. Certain infested non-food items may be treated with insecticide. Also treat pantry areas, drawers, shelves, cracks, crevices, moldings, and other infested areas, using a residual or contact spray of: 1.1% Baygon, 1–2% DDVP, 0.5% Dursban or diazinon, 0.25% Ficam, 1% lindane, or 2% malathion.

## CLOTHES MOTHS (Webbing Clothes Moths)

*Phylum: Arthropoda       Class: Insecta       Order: Lepidoptera*

This is a common straw-colored moth without dark spots on the wings. The clothes moth, or webbing clothes moth, infests fibers, hair, hair masses, woolens, silk, felt, and similar materials on which the larvae feed, often causing serious damage. When fully grown, the larva forms a cocoon in which it pupates.

### Control

See: *Casemaking Moths* (Control) in this section.

## CLOVER MITES (Brown Mites)

*Phylum: Arthropoda       Class: Arachnida       Order: Acarina*

Clover (brown) mites are plant feeders, and do not bite man. Nevertheless, these mites may invade houses in great numbers. The clover mite is a serious pest of plants, grasses, deciduous trees, and conifers, including violet, zinnia, apple, apricot, tomato, sycamore, sweet pea, strawberry, poplar, plum, pear, peach, ivy, iris, dandelion, cherry, beans, and clover.

### Control

Inside, use a vacuum cleaner to remove all visible mites from surfaces. Follow up with a residual spray to all infested surfaces. Another method is to tightly close the house, remove all inhabitants and pets, and release two or three aerosol insecticidal or miticidal bombs. Allow these bombs to work for three to four hours before venting and re-occupying the house. The following chemicals are recommended for interior spraying: 0.5% lindane, diazinon, or Dursban; or 2% malathion.

Outdoors, thoroughly apply a residual or contact spray to all affected plants, trees, grasses, lawn, outer walls and foundations of houses and buildings, steps, porches, crawl spaces, door facings, and other affected areas. Any of the following insecticides and miticides may be used: 1–2% malathion; 0.1% chlorobenzilate; 0.5% diazinon; 1–2% ethion, Kelthane, Ovex, Omite, Aramite, Plictran, Mitox, Tedion, Vendex, Pentac, Galecron, Fundal, fenson, Trithion, or Thiodan.

# CLUSTER FLIES

*Phylum: Arthropoda*     *Class: Insecta*     *Order: Diptera*

The cluster fly is dark gray with light-colored patches on the abdomen, and is slightly larger then the housefly. Cluster flies normally live outdoors, where they frequent flowers, fruits, plants, and other vegetation. When cool weather approaches, however, these flies enter houses where they may spend the winter, and may collect like a swarm of bees. They tend to hide in nooks, corners, clothing, closets, behind curtains, valances, pictures, and furniture, and other such concealed places. These flies are sluggish, and may fall suddenly to the floor, tables, and furniture without apparent reason. When crushed they emit a buckwheat honey odor. Cluster fly larvae parasitize earthworms in the soil.

## Control

Indoors, apply a residual or space spray, or release several aerosol bombs, containing any of the following insecticides: 1% ronnel, pyrethrins, Gardona, SBP-1382®, or DDVP; 0.5% Dibrom, Entex, or diazinon; or 2% malathion.

Outdoors, treat infested surfaces and/or plants and vegetation with the same residual or space spray. Also, if the soil nearby in flowerbeds, seedbeds, compost areas, etc. contains large numbers of earthworms, it may be necessary to eliminate the worms, since cluster fly larvae feed on them. However, earthworms generally are beneficial to the soil, and you should weigh this factor heavily before deciding to poison them.

# COCKROACHES

*Phylum: Arthropoda*     *Class: Insecta*     *Order: Orthoptera*

Cockroaches are ubiquitous! They are the most abundant, persistent, and bothersome pests in many houses, restaurants, and other buildings throughout the world. Indeed, cockroaches are possibly the filthiest and most repulsive of all pests because of their abhorrent living habits and the nauseous odor they produce. In addition, cockroaches may carry the disease organisms that cause various gastrointestinal disorders, such as cholera, typhoid, diarrhea, and food poisoning.

Often called "waterbugs," four different types of cockroaches cause most infestations in houses and restaurants: the German cockroach (croton bug), the American cockroach, the Oriental cockroach, and the brown-banded cockroach. Although these roaches differ somewhat in appearance, varying in color from tan to chestnut brown to black, all are flattened on top and tend to run rapidly across floors, walls, and other surfaces. They rarely fly. At rest, the head tends to flex downward and backward, so that it is almost hidden under the pronotum (thorax). Cockroaches are nocturnal, feeding and foraging at night and hiding during the day in cracks, crevices, walls, doorframes, baseboards, furniture, cupboards, pantries, steam tunnels, animal quarters, basements, and sewers. When a light is turned on suddenly, cockroaches may be seen scurrying to safety, often in swarms.

Cockroaches discharge fluids that impart a sickly, musky odor to rooms, foods, dishes, and other infested items. Cockroaches eat carbohydrates and practically all human foods, as well as glue, paste, book covers and pages, feces, and sputum.

The cockroach life cycle consists of three stages: egg, nymph, and adult. Nymphs molt several times before becoming adults, the time for this varying from a few weeks to several months, depending on the species. The major cockroach types are discussed below.

*German Cockroaches.* The most persistent and troublesome of cockroaches in most houses and restaurants, the German cockroach is about one-half of an inch or less in length, grayish to brownish in color, with two black longitudinal bars on the thorax behind the head. This roach is particularly fond of kitchens and bathrooms. The developmental period from egg to adult is about two to three months.

*Brown-Banded Cockroaches.* These cockroaches are also small, bothersome, and almost as common as the German cockroach. About one-half of an inch or less in length, the brown-banded cockroach has two brownish to yellowish transverse bands on the wings. Unlike the German cockroach, which has two black bars on the thorax, the brown-banded cockroach has a single dark area on its thorax. Nymphs have transverse yellowish-brownish bands on the abdomen and reddish-to-brown markings on the thorax.

*Oriental Cockroaches.* These roaches are practically black, and measure three-fourths to one and one-fourth inches long. Their wings do not completely cover the body. They frequent sewers and other filthy areas, and also live "in the wild" in such areas as outbuildings, yards, and basements. These roaches emit a strong, disagreeable odor. A cold-weather species, the Oriental cockroach is found most commonly in the northern United States. These roaches usually require a year or more to reach adulthood.

*American Cockroaches.* The American cockroach is a large species, measuring up to one and one-half inches or more in length. It is a dark, reddish brown, with a yellowish tinge on the thorax. Commonly found in buildings, this large roach prefers a warm, humid environment. Therefore, it frequents sewers, boiler rooms, basements, kitchens, and other favorable areas. Outside, the American cockroach can be found in wood piles, treeholes, trash piles, compost heaps, etc. These roaches reach adulthood in about a year.

### Control

Effective cockroach control requires good sanitation as well as proper insecticides. Never leave food exposed, keep garbage containers tightly closed, and clean up leftover bits of dog and cat food. In addition, you should fix any leaking water pipes and screen all sewer openings.

Chemical control of cockroaches usually requires one or more of the following pesticide formulations: oil-based sprays, water emulsion sprays, dusts, and baits. Generally, both a contact and a residual spray should be applied, using a fan or cone nozzle for surfaces and a pin-stream nozzle for cracks and crevices. Thoroughly treat all infested surfaces, cracks, crevices, holes, voids, cabinets, drawers, baseboards, bathrooms, door frames, etc.

Dusts are useful for cracks, crevices, the undersides of appliances, wall voids, subfloors, and other hard-to-reach areas. Use a light application only, since a heavy application tends to repel, rather than kill, cockroaches. While dusts tend to last longer than sprays, they become ineffective when damp, the exception being the wettable powders. Dusts should be applied with a pressurized applicator, bellows duster, or simple hand duster.

Baits are clean and easy to use, but they usually require a long time period to achieve good control. For the quickest and best results in controlling cockroaches, use sprays, dusts, and baits simultaneously. Be sure, however, that your baits contain an effective food lure such as peanut butter or syrup.

Recommended sprays are: 3–4% malathion; 1% diazinon; 2% ronnel; 0.5% Dursban; 1% Baygon; 0.25% Ficam W; or Roach-Prufe or Knox-Out (following label instructions). Suggested dusts are: 2% Diazinon D; 5–10% Sevin D; sodium flouride; silica aerogel; Roach-Prufe; Knox-Out; Ficam D; or Drione. Good baits contain 2% Baygon or 2% Kepone.

## CODLING MOTHS

*Phylum: Arthropoda*    *Class: Insecta*    *Order: Lepidoptera*

These are small, brownish to gray moths, often with bands or mottled marks on the wings. The codling moth attacks apples and other fruits. The adult moths appear in late spring and lay their eggs on tree leaves. The larvae (caterpillars) hatch and migrate to the young apples, chew into the fruit (usually through the bloom) and complete their development inside the fruit. Pupation occurs on the ground, under the tree bark, or in a similar concealed place. In the eastern states, a second generation of codling moths occurs in late summer, and the larvae spend the winter in cocoons under the bark of apple and other fruit trees.

### Control

In late spring, or at the first sign of adult moths or caterpillars, spray all fruit trees with a wettable powder or emulsion of: 2–3% Sevin, methoxychlor, Orthene, malathion, Ambush, Pounce, Lannate, or Pydrin. If a second generation of moths appears in late summer, it may be necessary to treat fruit trees again—especially to reduce or eliminate the larvae that otherwise will spend the winter and produce a new crop of moths the following spring.

## COLORADO POTATO BEETLES

*Phylum: Arthropoda*      *Class: Insecta*      *Order: Coleoptera*

This is a large yellow-and-black striped beetle that is a serious pest of potatoes throughout the United States and most of Europe.

### Control

Apply to affected potato plants the dust form of: 1–2% Sevin, Gardona, Guthion, malathion, or methoxychlor. Repeat this treatment as needed.

## CONENOSE BUGS

See: *Assassin Bugs* in this section.

## COOTIES

See: *Body Lice* in this section.

## CORN BORERS

*Phylum: Arthropoda*   *Class: Insecta*   *Order: Lepidoptera*

These caterpillars (borers) are the larvae of a moth that is yellowish-brown with darker markings, and with a wing-span of about one inch. There are two generations of corn borers per year, and these larvae live inside corn stalks and other plants which they may damage seriously.

### Control

Treat corn and other infested plants with a dust or residual spray of: 2–3% Sevin, malathion, methoxychlor, Orthene, Gardona, Guthion, Ambush, Pounce, Bolstar, Lannate, or Pydrin; or 1–2% Dursban or diazinon. Repeat this treatment as necessary.

## CORN EARWORMS

*Phylum: Arthropoda*   *Class: Insecta*   *Order: Lepidoptera*

Corn earworms are the larvae of medium-sized, heavy-bodied moths that generally are nocturnal (active at night). These moths may be attracted to lights at night and, thus become pests in this way, as well as by producing destructive larvae. The corn earworms feed on corn, tomatoes, cotton, and other plants, and sometimes are called *tomato fruitworms* or *cotton bollworms*. These larvae enter the corn ear through the silk and eat the kernels along the cob to the stalk. These larvae bore into tomato fruit, and into cotton bolls, and may do extensive damage to the crops discussed.

### Control

Apply the treatment recommended for *Corn Borers* in this section.

## CORN ROOTWORMS

*Phylum: Arthropoda*   *Class: Insecta*   *Order: Coleoptera*

Corn rootworms are the larvae of beetles. The adult beetle is a small, soft-bodied insect, usually yellowish with dark spots or stripes, and varying from a fraction to one-half of an inch long. The larvae (rootworms) are small, white, soft-bodied "worms" that feed on the roots and underground stems of corn, cucumbers, and other plants, to which they do serious damage.

### Control

Follow label directions on the insecticide container, and treat infested plants with any of the following chemicals: Furadan, Amaze, Dursban, ethoprop (Mocap), Dasanit, fonofos (Dyfonate), Thimet, or Counter.

## COTTON BOLLWORMS

See: *Corn Earworms* in this section.

## COWBIRDS

See: *Blackbirds* (Control) in this section.

## COWPEA WEEVILS

*Phylum: Arthropoda     Class: Insecta     Order: Coleoptera*

The cowpea weevil (pea weevil) lays its eggs on the pods of pea plants in the field. The larvae hatch out and bore into the pea seeds, often causing considerable damage to the crops.

### Control

Use the control methods recommended for the *bean weevil* in this section.

## CRAB LICE ("Crabs")

*Phylum: Arthropoda     Class: Insecta     Order: Anoplura*

Crab lice, commonly called "crabs," are small, grayish-white insects with the second and third pairs of legs enlarged, giving them a sea-crab appearance. These lice commonly occur on hairs in the pubic and anal areas of humans, but also may occur on the chest, in the armpits, and even in the eyebrows and eyelashes. Crab lice infesting the eyebrows feed in a localized area, causing small hemorrhages beneath the skin, thus giving it a bluish color. Crab lice eggs are glued to hair shafts. Three nymphal stages occur before adulthood, which takes about seventeen days. Crab lice usually are spread by sexual contact, but they also may be acquired from infested toilet seats, beds, and by close physical contact with infested persons. Although they suck blood, crab lice normally do not transmit disease.

### Control

Normally, human louse infestation is considered to be a medical problem and a physician is consulted. However, you can treat bodily infestations of lice, as well as other infested areas, quite successfully. In addition to regular bathing and changing of clothes, apply to hairy areas of the body any of the following chemicals: 1% lindane or malathion powder; or a 1% emulsion of lindane, 0.2% pyrethrins, or 12% benzyl benzoate. For eyebrow and/or eyelash infestation, an ophthalmic ointment containing 0.25% physostigmine is recommended. NOTE: For children under one year of age, consult a physician when lice infestation is detected.

For louse-infested restrooms, bathrooms, and other areas, wash toilets and seats with Lysol or a similar strong disinfectant.

### CRANE FLIES (Cling Flies)

*Phylum: Arthropoda      Class: Insecta      Order: Diptera*

Crane flies, also called cling flies, are large light-brown to caramel-colored insects that greatly resemble giant mosquitoes. They have extremely long legs that also make them resemble daddylonglegs with wings. Crane fly adults are about one inch or more in length. They occur mainly in damp areas with abundant vegetation.

Large numbers of crane flies frequently invade carports, patios, and homes, especially in the spring. When this occurs, they may be mistaken for "giant mosquitoes" to the great consternation of householders. These large flies do *not* bite humans, however, and are pests simply by their presence as they enter the home and cling to walls, ceilings, furniture, lamps, and other objects. In some cases, crane flies may feed on and damage cultivated plants or flowers.

### Control

Crane fly invasions of homes last only a few weeks, then these bothersome insects disappear. When present in large numbers, however, they are a nuisance as they cling to walls, ceilings, furniture, and other objects.

Crane flies inside are easily killed with a simple fly swat. Large numbers of these pests outside in carports or patios, or on flowers or plants, may be treated with a contact spray.

Recommended insecticides are: DDVP, dimetilan, Baytex, Dibrom, malathion, diazinon, SBP-1382®, ronnel, Rabon, Lethane, and Cygon.

## CROTON BUGS

See: *Cockroaches* (*German*) in this section.

## CRICKETS

*Phylum: Arthropoda*    *Class: Insecta*    *Order: Orthoptera*

There are several common species of crickets, but the house cricket is the most troublesome pest in this group. These crickets often enter houses in the summer and fall, and it may be nearly impossible to find exactly where or how they get inside. Once inside, crickets are equally hard to locate, and their chirping makes them nearly intolerable, especially at night. Look for them around or behind baseboards, in closets, beneath furniture and appliances, in and around fireplaces, and especially in soiled clothes or laundry. Crickets are quite fond of dirty socks, underwear, and other soiled laundry on which they like to chew.

### Control

First, try to locate the entry point(s) used by crickets to get inside, and close these off. Outside, thoroughly treat the ground, yard, and building foundation with a residual spray of: 0.5% diazinon, 1.5% Baygon, 0.5% Dursban, 0.5% Ficam, or 2% malathion or Sevin. The following dusts also are recommended: 5–10% Sevin, 2% diazinon, or 5% malathion.

## CUCUMBER BEETLES

See: *Corn Rootworms* in this section.

# CUTWORMS

*Phylum: Arthropoda*     *Class: Insecta*     *Order: Lepidoptera*

Cutworms are the larvae of a large moth group, which also produces the army worm and corn earworm. These moths are heavy-bodied, and many are attracted to lights at night. The larvae (cutworms) feed on the roots and shoots of various plants, and cut off the plant at ground level. Cutworms are nocturnal and hide under stones and in the soil during the day. They do extensive damage to crops and plants.

### Control

Thoroughly treat the lawn, turf, and the soil around plants with a residual or contact spray of: 1–2% Baygon, diazinon, or Sevin; 0.5% Dursban; or 2% trichlorofon. The following dusts may be used to treat plant parts above ground: 5–10% Sevin; 3–5% malathion; or 2% diazinon, Orthene, Ambush, Pounce, Lannate, Bolstar, or Pydrin.

# DADDYLONGLEGS (Harvestmen)

*Phylum: Arthropoda*     *Class: Arachnida*     *Order: Phalangida*

Daddylonglegs (harvestmen) may be confused with true spiders which they resemble somewhat. Daddylonglegs are arachnids, but *not* true spiders. They have eight very long, slender, arched legs that support a small body that is brownish in color. Daddylonglegs drink water frequently, and must have a ready supply available in or near areas they infest. Although they are harmless, daddylonglegs may become pests when they invade houses in large numbers.

### Control

Attempt to locate breeding areas and nests and remove these. Also, eliminate all open or standing water sources. If daddylonglegs are present in numbers great enough to warrant treatment, apply the methods and insecticides recommended for black widow and brown recluse spiders.

# DEERFLIES

*Phylum: Arthropoda*     *Class: Insecta*     *Order: Diptera*

Deerflies are similar to horseflies, but somewhat smaller. Both flies are common pests throughout many parts of the world. Deerflies are medium sized to large, stout-bodied, strong flying insects. The females are blood-suckers, and are serious pests of horses, livestock, and man. Most species of deerflies are slightly larger than the housefly, brown to black in color, with dark markings on the wings. They live near marshes, swamps, streams, and other water sources in which they breed. Some deerflies are believed to transmit the disease organisms of tularemia (rabbit fever), anthrax, and possibly other diseases.

### Control

Deerflies are difficult to control because they breed in marshes, swamps, streams, and other such areas—often far removed from where they pose problems. For temporary relief, however, the following space sprays are recommended: malathion, dichlorvos, pyrethrins, resmethrin, SBP-1382®, Gardona, Dibrom, trichlorfon, Cygon, Baytex, and diazinon.

## DOBSONFLIES

*Phylum: Arthropoda      Class: Insecta      Order: Neuroptera*

Dobsonflies are similar to alderflies, but are larger, measuring an inch or more in length. Usually found near water, dobsonflies are soft-bodied insects with a fluttery flight pattern. Some species are attracted to lights at night, and become pests during the warm months.

### Control

Dobsonflies usually are only minor pests, except when they swarm around lights at night. As with all water-breeding flies and insects, however, these pests may be controlled best by locating their breeding areas and treating these with any of the following chemicals: Abate, Altosid, diesel fuel, fuel oil, or crankcase oil.

Temporary relief against flying adults may be obtained by applying any of the following space sprays: DDVP, dimetilan, Baytex, Entex, Dibrom, malathion, diazinon, SBP-1382®, Lethane, Cygon, ronnel, or Gardona. Additionally, a residual spray of any insecticide listed should be applied to surfaces frequented by these insects. Also, one or more electronic bug killers installed around outside lights may be effective in controlling these pests.

## DOODLEBUGS

See: *Ant Lions* in this section.


## DRAGONFLIES

*Phylum: Arthropoda*    *Class: Insecta*    *Order: Odonata*

Dragonflies are rather large, harmless, flying insects, measuring from one to three and one half inches long. Most species are good fliers, and spend much of their time airborne, feeding on smaller flies, mosquitoes, midges, etc. which they catch on the wing. Dragonflies normally do not become pests, except occasionally during hot, dry weather, when they may swarm in great numbers in yards and around houses and buildings.

### Control

Control measures for dragonflies should not be necessary except in very rare cases, when they swarm in great numbers in yards and around houses. Even then, treatment is questionable, because the flies normally disappear in a few days. If, however, treatment is attempted, a tractor-mounted or portable engine-driven sprayer or fogger will be required to adequately dispense any of the following insecticides: DDVP, dimetilan, Baytex, Entex, Dibrom, malathion, diazinon, SBP-1382®, Lethane, Cygon, ronnel, or Gardona.


## DRAIN FLIES (Filter Flies)

*Phylum: Arthropoda*    *Class: Insecta*    *Order: Diptera*

These are very small, darkish-grayish, furry flies that are less than one-fourth of an inch long. Outside, they swarm around drains, sewers, and decomposing organic matter. Often called filter flies, these insects frequently appear (apparently from nowhere) inside the house, and hover around sinks and drains in the kitchen and bathroom. The adult female fly lays her eggs inside the drains of bathtubs, showers, and sinks, and the larvae develop on the soapy, hair-filled, organic crud that collects inside sinks and drains. The adult flies emerge from these drains, appearing to come from nowhere. Drain flies are harmless, and are pests only by their presence.

**Control**

Clean out bathtub, shower, and sink drains regularly and/or treat them with a strong disinfectant such as Lysol. Flying adults, both inside the house and outdoors, may be treated with one of the following aerosol or space sprays: DDVP, dimetilan, Baytex, Entex, Dibrom, malathion, diazinon, SBP-1382®, Lethane, Cygon, ronnel, or Gardona.


## DRUGSTORE BEETLES

*Phylum: Arthropoda      Class: Insecta      Order: Coleoptera*

Drugstore beetles are very common pests of pantries, stored foods, cereals, and various drugs. These beetles are cylindrical to oval, rather pubescent, and vary from a fraction to almost one-half of an inch in length.

**Control**

Drugstore beetles tend to occur in limited areas, such as a single kitchen cabinet, one shelf or drawer, or even one container of stored food or drugs. It is necessary to find this source and discard it. Check thoroughly: all cabinets, drawers, pantries, shelves, cracks and crevices, and inside the containers of cereals, beans, peas, flour, dried fruits, spices, and other such items. Remove all food, pots, pans, utensils, and shelf paper from drawers, cabinets, and pantries and treat these areas with a residual spray or dust. The following insecticides are recommended: 1–2% SBP-1382®, resmethrin, or pyrethrins; 1.5% Baygon; 2% DDVP or malathion; 0.5% Dursban or diazinon; or 0.25% Ficam.


## DRYWOOD TERMITES

*Phylum: Arthropoda      Class: Insecta      Order: Isoptera*

These termites live in wood above ground. They attack dry, sound wood, which they may damage extensively. Most drywood termite infestations occur in buildings; but furniture, utility poles, and stacked lumber also may be attacked. These termites bore out large chambers, cutting across the grain of the wood, and connect these chambers with many small tunnels. They may do extensive damage to buildings and other wood structures. The adult drywood termite is almost one-half of an inch long, cylindrical, and usually pale brown in color. Unlike the tunnels of powderpost beetles which are filled with sawdust and feces, drywood termite chambers and tunnels are clean.

### Control

A large infestation of drywood termites, particularly in the wood-work of buildings, is usually beyond the ability of the average person to treat effectively. In such cases, a professional pest control operator should be hired. Smaller and more limited infestations, however, may be treated by the average person.

To treat limited areas, drill one-half-inch diameter holes in larger timbers and smaller holes in smaller timbers. Use a pressurized hand duster or pressurized liquid sprayer to force either a dust or residual spray deeply into each drilled hole and, thus, throughout the entire tunnel network within the wood. Recommended dusts are: sodium fluorosilicate, silica aerogel, or Drie-die 67. Diazinon, Ficam, Dursban, and Baygon dusts probably will work equally well. A recommended liquid insecticide is 5% PCP (pentachlorophenol).

To prevent drywood termite attack, use a paint brush or fan-nozzle sprayer to apply a film of boric acid or Woodtreat-TC to all susceptible wood surfaces.

## EARTHWORMS

*Phylum: Annelida     Class: Chaetopoda     Order: Megadrili*

Normally earthworms in the soil are beneficial because they constantly mix and enrich the dirt. Occasionally, however, they may become too plentiful. If this occurs in turfs and lawns, the worms may loosen the grass roots, and also render the soil lumpy and uneven.

### Control

Any of the following treatments are recommended for earthworm control: mercuric chloride (2–3 ounces in 50 gallons of water per 1000 square feet of soil); ammonium sulfate (3 lbs. per 1000 square feet of soil); lead arsenate (1 lb. per 2 gallons of water, applied 20 gallons per 1000 square feet of soil; or 2–5% diazinon, trichlorfon, Dursban, ethoprop, Sevin, or Aspon. Mesurol, methiocarb, and Zectran also may be effective against earthworms.

# EARWIGS

*Phylum: Arthropoda    Class: Insecta    Order: Dermaptera*

Earwigs are easily identified by the prominent pair of forceps at the tip of the abdomen. They are medium sized, elongated and flattened insects with short wings. Normally earwigs live outside, hiding in debris, under bark, and in other concealed places during the day, and feeding on plants at night. Some species, however, do invade houses in great numbers. Two common earwig species that occur inside the house are the European earwig, which is about five-eighths of an inch long and reddish-brown in color; and the ring legged earwig, which is slightly over one-half an inch long, dark brown to shiny jet black, with yellow-brown legs striped with dark cross bands. The ring legged earwig also lives in greenhouses. This earwig is attracted to light and produces a strong odor when crushed.

## Control

Earwig control should begin outside and treatment inside is merely supplemental. Thoroughly apply a residual spray on the ground, in trash and debris piles, around the perimeter of the house or building, on the foundation, in crawl spaces, flowerbeds, and on turfs and lawns.

Inside, treat cabinets (both inner and outer surfaces), baseboards, cracks and crevices, and other concealed areas. The following sprays are recommended: 2% malathion or Sevin; 1.5% Baygon or diazinon; 0.5% Dursban; or 0.25% Ficam. 2% Baygon bait also is recommended. Finally, 10% diazinon or 10% Sevin granules can be used to treat outside ground areas.

# ELM BARK BEETLES

*Phylum: Arthropoda    Class: Insecta    Order; Coleoptera*

These are small, cylindrical beetles, rarely over one-half of an inch long, and usually brownish to black in color. Bark beetles live beneath the bark of trees, and eat into the wood. Most bark beetles attack only evergreen trees, but a few species, including the elm bark beetle, attack other trees. These pests often tunnel extensively under the bark, and may cause the tree to die. Moreover, some species of bark beetles transmit the deadly Dutch elm disease, which has killed thousands of elm trees.

**Control**

Treat all infested trees with a heavy residual spray of: 2% Sevin, diazinon, Baygon, Dursban, or lindane. Repeat this treatment as needed.

## ELM LEAF BEETLES

*Phylum: Arthropoda    Class: Insecta    Order: Coleoptera*

Adult elm leaf beetles vary from yellow to dull green in color, and are about one-fourth of an inch long. These beetles infest elm trees and make them susceptible to diseases, such as Dutch elm disease. These beetles may invade houses in great numbers, especially in the fall. They may spend the winter inside, especially in attics or within walls, and emerge again in the spring. New larvae are black and resemble slugs; full-grown larvae, however, are about one-half of an inch long, dull yellow with black head, legs, and hairs, and a pair of black stripes along the back. In some areas of the country, there are two generations of elm leaf beetles per year.

**Control**

Outside, thoroughly treat infested trees with a contact or residual spray of: 1–2% Sevin, malathion, diazinon, or lindane. Inside, treat areas where beetles and larvae occur with a residual or contact spray of: 1–2% Sevin or malathion; 1.5% diazinon or Baygon; 0.5% Dursban; or 0.25% Ficam.

## ENGRAVER BEETLES

*Phylum: Arthropoda    Class: Insecta    Order: Coleoptera*

These are small, cylindrical, brownish to black beetles, rarely over one-third of an inch long. They are closely related to and resemble the other bark beetles. The term "engraver beetle" comes from the characteristic engraver-pattern tunneling done by these beetles just beneath the bark of trees. Larvae pupate beneath the bark, then bore small holes in the bark through which they emerge. These holes in the tree bark resemble a shotgun spray pattern.

**Control**

Apply the same control measures recommended for *Elm Bark Beetles* in this section.

# EUROPEAN CORN BORERS

*Phylum: Arthropoda    Class: Insecta    Order: Lepidoptera*

European corn borers are the larvae of large, conspicuously colored moths that are yellowish-brown with darker markings. The larvae (borers) live within the corn stalks and other plants, and frequently cause extensive damage to the host plants. There are one to two generations of corn borers each year, with the last generation wintering in the larval form.

## Control

Infested corn stalks and other plants should be burned, shredded, or somehow destroyed in the fall to eliminate corn borers that may over-winter and produce a new infestation the following year.

During the growing season, treat infested corn and other plants according to container label directions, using any of the following insecticides: EPN, ethoprop, Furadan, Dasanit, Thimet, or Counter.

# FABRIC PESTS

The most common fabric pests are: clothes moths, webbing clothes moths, casemaking moths, carpet beetles, and furniture beetles. These pests attack fabrics and cloth materials, often causing expensive damage to clothes, carpets, furniture upholstery, piano felt, blankets, furs, hairy materials, and woolens. Each of these pests is covered individually, in alphabetical order, in this section.

# FACE FLIES

*Phylum: Arthropoda    Class: Insecta    Order: Diptera*

Face flies are so named because they characteristically swarm around the faces of cattle and other animals. These flies may enter houses and hibernate within the walls, emerging in the spring or on warm winter days.

## Control

Begin by fly-proofing the infested house or building.

Adult flies inside the house may be killed with any of the following aerosols or space sprays: DDVP, Entex, Gardona, pyrethrins, SBP-

1382®, trichlorofon, Dibrom, resmethrin, or ronnel. Killing only the visible adult flies will not eliminate this problem, however, since the flies constantly emerge from within the wall voids and other concealed areas. Effective control of face flies inside the house requires that an insecticidal dust be injected under pressure into wall voids and other hiding places. Use the dust form of any insecticide listed above, or use diazinon and malathion.

## FALL WEBWORMS

*Phylum: Arthropoda    Class: Insecta    Order: Lepidoptera*

Fall webworms are the larvae of a tiger moth. These larvae (webworms) feed on trees and shrubs, often causing serious damage. These worms build large webs, often enclosing an entire limb or branch, and feed within this giant web. These webs are common on various types of trees and shurbs, especially in the spring and fall.

### Control

See: *Caterpillars* (Control) in this section.

## FILTER FLIES

See: *Drain Flies* in this section.

## FIRE ANTS

*Phylum: Arthropoda    Class: Insecta    Order: Hymenoptera*

Several species of fire ants occur in the United States. These ants vary somewhat in size and color. The southern fire ant is a clear reddish-to-yellow color, and varies from a smaller fraction to about one-fourth of an inch long. The imported fire ant, another common species, is about the same size, but is reddish to blackish in color. Fire ants are very active, and often attack young animals. They inflict a painful sting to both animals and humans, often causing a severe reaction in allergic persons.

Some fire ants build their nests in loose soil, with many craters scattered over an approximate two to four square foot area. Openings to these nests usually are found under stones and boards, near grass tufts, in cracks within concrete, and beneath houses and buildings. Nests also may occur within the woodwork or masonry of houses and other buildings.

Other fire ant species are important agricultural pests, building their nests and earthen mounds in open fields, gardens, pastures, etc. where they seriously damage crops and plants.

### Control

Large infestations of fire ants in fields, pastures, and other extensive areas are difficult to control without special large-scale pesticide application equipment. However, for limited infestations in yards, turfs, gardens, sidewalks, and in and around buildings, the following measures are recommended: (1) attempt to locate the nest openings and/or earthen mounds; use a pin-stream nozzle and thoroughly spray the nest openings with any insecticide listed below; (2) if fire ants occur in or near the house or other building, attempt to locate the nesting area and treat it with a residual spray or dust, using any insecticide listed below; (3) while control will take longer, the following ant baits may be used: Mirex, Kepone, or Baygon; (4) finally, apply a 3-foot residual spray barrier, or scatter bait granules, around the entire perimeter of infested houses and buildings.

Recommended residual and contact sprays are: 1–2% Baygon, diazinon, or malathion; 0.5% Dursban, lindane, or Ficam; or 3% bioresmethrin, methoxychlor, or dichlorvos.

## FIREBRATS (Bristletails)

*Phylum: Arthropoda     Class: Insecta     Order: Thysanura*

Firebrats are very similar to silverfish, except for color and habitat. Firebrats are slightly less than one inch long, are tan to brown in color, and have three long tails at the tip of the abdomen, hence the term "bristletails." Firebrats are common pests in homes and restaurants, particularly around furnaces, boilers, steam pipes, stoves, fireplaces, and other warm areas. They are active, fast-running insects that feed on carbohydrates, wallpaper paste, starched cloth, book bindings, and stored food. They may become serious pests unless controlled.

### Control

Apply a residual spray to all possible hiding areas, especially: cracks and crevices in basements, cabinets, cupboards, and closets; behind baseboards and wooden moldings; and around water and steam pipes, boilers, heaters, stoves, and other warm areas. Dusts under pressure should be applied to wall voids, crawl spaces, subfloors, and other inaccessible areas.

Recommended sprays are: 1.5% Baygon or diazinon; 0.25% Ficam; 1–2% pyrethrins; or 0.5% Dursban. Recommended dusts are: 2% diazinon, silica gel, Ficam, or malathion.

To kill visible firebrats, use a space or contact spray of: 1–2% DDVP, pyrethrins, or SBP-1382®.

Recommended bait: Kepone pellets.

## FISHFLIES

*Phylum: Arthropoda*    *Class: Insecta*    *Order: Neuroptera*

Fishflies are very similar to alderflies and dobsonflies, lying between these two species in size. Fishflies are slightly larger than alderflies, but a bit smaller than dobsonflies. Fishflies are soft-bodied, fluttery flying insects that breed in water. They may be attracted to lights at night and, thus, become pests—especially during the warm months.

### Control

See: *Alderflies* and *Dobsonflies* (Control) in this section.

## FLAT GRAIN BEETLES

*Phylum: Arthropoda*    *Class: Insecta*    *Order: Coleoptera*

The flat grain beetle is a very small, flattened, oblong, reddish-brown insect about one-sixteenth of an inch long. It is a common pest of grain, cereals, and stored food, and may live in kitchen cabinets, cupboards, and pantries.

### Control

Locate all infested food materials and discard them. Remove all dishes, pans, utensils, and shelf paper from cabinets, drawers, pantries, and cupboards. Treat these areas with a residual spray or dust, paying close attention to all cracks and crevices, and other hiding places used by these very small beetles. Recommended residual sprays and dusts are: 1.1% Baygon; 1% DDVP; 0.5% Dursban or diazinon; 1% lindane; 0.25% Ficam; drione dust; or 2–3% malathion.

# FLEAS

*Phylum: Arthropoda*    *Class: Insecta*    *Order: Siphonaptera*

Fleas are very common pests of man and animals throughout the world. They are small, wingless insects, varying from barely visible to one-third of an inch long, but averaging about one-sixth of an inch in length. Fleas pierce the skin of their hosts and suck blood. Thus, some species of fleas can, and do, transmit disease. Most species of fleas infest small mammals such as rabbits, rats, moles, bats, and gophers. But some species infest larger animals and birds, and at least a few species attack humans.

The flea life-cycle involves complete metamorphosis: egg, larva, pupa, and adult. This life-cycle varies in length, depending upon temperature, humidity, availability of food, and other factors. Under normal conditions, however, a generation of fleas is produced in about two to three weeks. Flea eggs may be visible as small, rounded, bright-colored objects in furniture, upholstery, carpets, bedding, and in cracks and crevices. Eggs laid on an animal simply drop off, incubate, and hatch out larvae. Flea larvae are small, very active, wormlike creatures that may be spotted in the areas mentioned.

The more common flea species are discussed in more detail below.

***Cat and Dog Fleas.*** These very common fleas are small, dark insects that leap or jump great distances. Cat and dog fleas are almost identical, and must be distinguished by an entomologist. Cat fleas, however, are more common as pests of man. Both types of fleas infest cats, dogs, humans, and many other animals, and both types are intermediate hosts of the dog tapeworm. Thus, children, when playing with flea-infested cats and dogs, often ingest fleas accidentally and become infested with tapeworms.

Cat and dog fleas often become serious pests inside the house, where they live in carpets, furniture, bedding, and similar places. When an animal is not present from which to suck blood, these fleas will attack humans, often viciously and in great numbers, especially when the house has been vacated for a few days. Upon returning, humans may face a sudden attack by a large group of hungry fleas which bite viciously.

A flea bite typically produces a small, hard, red knot or welt that itches and may even bleed. The flea bite produces a single puncture mark in the center of the knot, thus differing from ant and spider bites which leave two marks. Cat and dog fleas normally do *not* transmit disease.

***Sticktight Flea.*** This flea occurs primarily in the south and southwestern United States. Essentially it is a pest of poultry, but also will attack man and other animals. The female sticktight flea attaches firmly to the host, often causing severe sores and ulcers on the head, neck, and ears of fowl and animals, and almost anywhere on the body of a human. This flea can transmit a form of typhus fever to humans.

*Mouse Flea.* This flea occurs on rats and mice, particularly in the Gulf States and in California. Heavy mice and/or rat infestations inside a house or building may cause this flea to become a problem.

*Human Flea.* This flea occurs throughout warmer parts of the world. it is the most important flea species attacking man along the Pacific Coast, but also causes pest problems in the midwest and southern states. Human flea bites cause dermatitis and allergic reactions of the skin. Infestations of these fleas commonly occur in barns, barnyards, hogpens, and other areas adjacent to human and animal habitats. Once a colony of human fleas becomes established in hogpens and other animal quarters, these pests may remain for weeks or months after the animals are gone.

*Squirrel Flea.* This flea may be found wherever gray squirrels live. Squirrel nests in attics and other areas may cause this flea to become a serious household pest.

*Chigoe Flea.* This small flea occurs in tropical and subtropical parts of North America, South America, the West Indies, and Africa. The female flea burrows into the skin of her hosts, frequently between the toes or under the toenails. The embedded flea gradually becomes engorged with blood and eggs, often swelling to the size of a pea, and causing excruciating pain to the victim. Inflammation and ulcers may result from these embedded fleas, and occasionally the secondary infection causes tetanus or gangrene.

### Control

Flea control, both inside and outdoors, should begin with cats, dogs, and all other infested animals. Following this, both the inside and outside premises should be treated.

*Fleas on Cats and Puppies.* Cats and puppies can be treated with the following dusts: 1% pyrethrins, 0.2% pyrethrins-plus-synergist (piperonyl butoxide); 1% rotenone; or 2-5% malathion or Sevin. Recommended sprays are: 0.5% malathion or Sevin. Rub the dust well into the fur and hair, covering the entire animal without getting it into the eyes and nostrils. Flea collars also may be used on both cats and dogs, but these generally are not as effective as the other methods discussed.

*Fleas on Dogs.* Recommended dusts are: 1% pyrethrins, 0.2% pyrethrins-plus-synergist; 1% rotenone or lindane; 0.5% Co-Ral; 4-5% malathion; or 2-5% Sevin. Recommended sprays are: 0.5% malathion or Sevin.

*Outdoor Premise Treatment.* Thoroughly treat animal pens, kennels, houses, grounds, yard, turf, and grass with a dust or residual spray of: 1%

lindane or diazinon; 2% Baytex; 1–2% malathion, ronnel, or trichlorofon; 2–5% Sevin; or 1% Baygon.

*Inside Premise Treatment.* Flea-infested cats or dogs allowed inside often establish a flea colony inside the house. These fleas tend to live in carpets, rugs, furniture, upholstery, bedding, and similar material. They will attack human inhabitants, and may become intolerable. Vacuum all carpets, rugs, and upholstery, and thoroughly clean bedding at least once per day. Discard the vacuum bag each time. Close off the house tightly, remove all living inhabitants, and release three or four aerosol insecticide bombs that specifically kill fleas. Allow these bombs to work for three to four hours before venting and re-occupying the house.

Usually this treatment, combined with daily vacuuming, will eliminate an infestation of fleas in ten to fourteen days, provided animals do not bring new fleas inside. However, if the infestation is severe and fleas persist, it may be necessary to treat infested areas with one of the following water-based sprays or dusts: 0.25% Ficam; 1% ronnel, DDVP, or pyrethrum; 2% malathion or Sevin; Whitmire Prescription Treatment #585; or 0.5% SBP-1382®.

## FLESH FLIES

*Phylum: Arthropoda     Class: Insecta     Order: Diptera*

Flesh flies are similar to blowflies, but are blackish-grayish with gray stripes on the thorax, rather than metallic colored. Adult flesh flies are common pests, and the larvae (maggots) feed on animal matter. Adult flesh flies have a reddish-orange spot at the tip of the abdomen.

### Control

See: *Blowflies* (Control) in this section.

## FLYING INSECTS

There are many types of flying insects that are pests of man. Often they seriously effect his comfort and health. Generally, flying insects are: flies (many types), mosquitoes, midges, gnats, black flies, alderflies, fishflies, dragonflies, wasps, bees, hornets, and other less common insects.

### Control

In many cases, it is difficult to treat effectively against flying insect pests because their nesting and breeding areas often are located some distance away, or are too widespread and extensive to treat effectively. In this case, temporary relief may be gained by using a space spray to kill flying adults and a residual spray to treat landing surfaces used by these insects. In a few cases, particularly with flies, poisoned baits may be used with some success.

The following space and residual sprays are recommended for flying insects: DDVP, dimetilan, Baytex, Entex, Dibrom (naled), malathion, diazinon, resmethrin, thanite, Lethane, SBP-1382®, dimethoate, Cygon, ronnel, and Gardona. Sevin is effective against bees, but not against flies.

## FLOUR BEETLES

*Phylum: Arthropoda      Class: Insecta      Order: Coleoptera*

Generally, flour beetle adults are black or dark brown, and slightly over one-half of an inch long. Some species, however may be smaller—about one-quarter of an inch long. These beetles may become pests inside the house where they infest flour, cornmeal, and other stored food. The larvae of these beetles are called *mealworms.*

### Control

Discard all infested food materials. Remove all pans, dishes, utensils, and shelf paper from all cabinets, drawers, pantries, and cupboards. Treat these areas with a dust or residual spray, using any insecticide listed: 2% malathion, 1.1% Baygon, 0.5% diazinon or Dursban, 0.25% Ficam, or 1–2% DDVP.

## FLOWER BEETLES

*Phylum: Arthropoda      Class: Insecta      Order: Coleoptera*

Flower beetle species found in the United States are small- to medium-sized insects, but some African species are four to five inches long. Some adult flower beetles feed on pollen, while others feed on decaying wood and live under bark. The larvae of these beetles live in the soil, where they feed on and damage the roots of various flowers and plants. One species of flower beetle, the green June beetle, feeds on grapes, fruits, foliage, and young corn. The larvae of these beetles also feed on plant roots in the soil, often causing damage. The green June beetle is common throughout the southeastern United States.

**Control**

Affected plants should be treated with a water-based contact or residual spray, or a dust, using any of the listed insecticides: 1-2% malathion, Sevin, methoxychlor, or lindane; 2-3% Guthion or Gardona; or 1.5% Baygon, Dursban, or diazinon. For plant or flower rootworms within the soil, follow container label directions and apply any of the following agents: Furadan, Amaze, Dursban, ethoprop, Dasanit, fonofos, Thimet, or Counter.

## FUNGI

Fungi are living plants that attack wood, causing damage and decay. Fungi are a particular problem in damp, humid areas of the world because moisture is always necessary for them to thrive. Permanent control of fungi thus requires proper drainage, permanent ventilation, and drying of affected wood and timbers.

Fungus infestations of wood can be treated with wood preservatives. PCP (pentachlorophenol) in oil, or Woodtreat-TC should be brushed on heavily with a paint brush, or sprayed on with a pressurized sprayer, using a fan nozzle.

## FURNITURE BEETLES

*Phylum: Arthropoda      Class: Insecta      Order: Coleoptera*

Furniture beetle adults are small, red-to-blackish brown insects, usually less than one-third of an inch long. These beetles attack wooden furniture and wooden structures within buildings, with infestations commonly occurring in crawl spaces, basements, cabinets, woodwork, pantries, and cupboards. Small round holes on the surface of wood, with very fine sawdust or powder sifting out, is evidence of these beetles.

**Control**

If hardwood flooring is involved, it is also necessary to treat the subflooring beneath. Apply a residual spray, using a fan or cone nozzle, to all infested surfaces. When treating finished wood (flooring, furniture, cabinets, etc.) use an oil solution of insecticide. However, first you should treat a small area to see whether spotting or staining will occur. If so, you should switch to a different oil solvent.

Recommended insecticides are: 5% PCP or 0.5% lindane. If adult beetles are visible, the following chemicals are recommended to kill them: 0.5% DDVP, SBP-1382®, or pyrethrins.

## FURNITURE CARPET BEETLES

*Phylum: Arthropoda*    *Class: Insecta*    *Order: Coleoptera*

Adult furniture carpet beetles are small, rounded, and blackish, with a sprinkling of yellow and white markings on the back and heavily coated yellow scales on the legs. These pests often attack upholstered furniture where they feed on the hair, padding, feathers, and woolen material inside. These beetles also attack woolens, carpets, rugs, fur, bristles, silk, and other materials. They also feed on fibers of linen, cotton, rayon, and jute which are stained with feces. These beetles are serious pests, and may do considerable damage to fabric and fiber materials.

### Control

Control procedures are essentially the same as those recommended for *Buffalo Bugs* and *Casemaking Moths*. Thoroughly vacuum and clean carpets, rugs, and furniture upholstery. Also check and air, launder, or dry clean all fiber materials, such as linen, cotton, rayon, silk, and jute. Discard all old stored garments, bedding, carpets, rugs, and pieces of carpet and woolen materials. Vacuum thoroughly around baseboards, air ducts and vents, and beneath furniture. When chemical treatment is required for heavy or persistent infestations, use any of the insecticides listed for *Buffalo Bugs* and *Casemaking Moths* control. However, check the label directions and use a solvent that will not stain or spot fabrics.

## GALLINIPPERS

*Phylum: Arthropoda*    *Class: Insecta*    *Order: Diptera*

The gallinipper is a very large species of *Psorophora* mosquito, common in Mexico and parts of the United States northward to Canada. Adults are yellowish-brown with shaggy legs, and are vicious biters, both during the day and at night.

### Control

See: *Mosquitoes* (Control) in this section.

## GNATS

*Phylum: Arthropoda*    *Class: Insecta*    *Order: Diptera*

There are many types of gnats, varying greatly in size, color, and pest activity. Small black gnats, one of the most bothersome types, are discussed under black flies/black gnats in this section. Other gnats, however, are much larger, slender insects that resemble mosquitoes. These gnats usually exist where there is an abundance of decaying vegetation, and may become pests in damp, moist cellars and similar places.

### Control

Gnats cannot be controlled completely by a single person. Control measures must, therefore, aim for temporary relief. Use an insect repellent for protection when gnats swarm in large numbers. An aerosol or space spray is effective in enclosed areas such as buildings, rooms, basements, and cellars. Use any insecticide listed for *Flying Insects* in this section.

## GOPHERS (Pocket Gophers)

*Phylum: Chordata      Class: Mammalia      Order: Rodentia*

Gophers are buffy brown to light gray mammals, with gray to white underbellies. The eyes and ears are small and the enlarged forefeet bear long claws used for digging. Gophers breed throughout the year, producing one to three litters which average five to six young each. These pests are detected easily by their burrows where mounds of fresh dirt are pushed to the surface. Mounds are crescent-shaped, and the plugged opening is not in the center in most cases. Gophers are solitary animals and will defend their burrows against encroachment by other gophers.

Gophers frequently invade lawns, gardens, turfs, golf courses, and other areas, particularly in the spring and summer when the young ones leave the maternal burrows. These animals eat the leaves and stems of plants, the bark of trees, and the roots and bulbs of various plants and vegetables. They may become serious pests.

### Control

If at all feasible, fumigation of burrows is the quickest and most efficient way to eliminate gopher infestations. Carbon monoxide, hosed from your automobile exhaust pipe into gopher burrows, is quite effective as a fumigant. Also, liquid PCP (pentachlorophenol), carbon tetrachloride, or methyl bromide are good fumigants for burrows.

Another recommended control method is the use of poison baits. Root vegetables (carrots, sweet potatoes, turnips, etc.) sliced and laced with strychnine in the center make good baits. Add about one half of an ounce of powdered strychnine sulfate to about four quarts or less of sliced or cubed vegetables. Use a probe to make small holes in the gopher burrows, and drop in baited vegetables at about eight foot intervals. Close the holes with loose dirt. Additionally, two commercial gopher baits are recommended: Gophacide and Gopha-Rid.

Finally, gophers may be trapped and killed or otherwise disposed of individually. Macabe and box traps are recommended for best results.

## GRACKLES

See: *Blackbirds* (Control) in this section.

## GRAIN BEETLES

See: *Saw-Toothed Grain Beetles, Cigarette Beetles, Drugstore Beetles, Cadelles, Flat Grain Beetles,* and *Red Flour Beetles* in this section.

## GRAIN MITES

See: *Cheese Mites* in this section.

## GRAIN WEEVILS

*Phylum: Arthropoda    Class: Insecta    Order: Coleoptera*

These are very important pests of stored wheat, corn, rice, barley, and other grains. They also may infest stored foods inside the house. Grain weevils are small, brownish insects, about one-sixth of an inch long, with long snouts.

### Control

Large bins of stored grain infested with these weevils will require fumigation for effective control. Recommended fumigants are phosphine and acrylonitrile. Follow label directions. NOTE: In most cases large scale fumigation must be done only by licensed pest control operators.

For small-scale weevil infestations inside the house, however, check stored grains, foods, cereals, oats, spices, etc. and discard any that contain weevils. Remove all pots, pans, dishes, and shelf paper from all cabinets, drawers, cupboards, and pantries. Treat these areas with a residual spray or dust, paying close attention to all cracks and crevices and other hiding areas.

The following sprays and dusts are recommended: resmethrin, pyrethrins, SBP-1382®, and DDVP contact or space sprays; 1.1% Baygon, 0.5% Dursban or diazinon, 0.25% Ficam, or 2% malathion spray or dust.

## GRASS BUGS

*Phylum: Arthropoda      Class: Insecta      Order: Hemiptera*

The grass bug is a very small species, only a fraction of an inch long, with a darkish body and whitish, translucent wings that give it a grayish appearance. This bug feeds on grasses, and may reproduce in great numbers in vacant lots, fields, along fence rows, and other grassy areas. When the grass becomes dry, however, or is mowed or burned, the adult and nymph grass bugs migrate. At this time, they may swarm in great numbers on fences, the sides of houses, and invade houses and buildings.

### Control

For ground treatment, use a residual spray or dust. For fence and outside building treatment, use a residual spray only, unless the bugs are visible in great numbers. In this case, a contact spray is recommended. Thoroughly treat the sides, foundation, outside walls, porches, steps, doors, and windows of buildings. The following insecticides are recommended: Baygon, lindane, malathion, diazinon, Dursban, Sevin, and methoxychlor.

For indoor infestations, use a residual spray or contact spray. Also, aerosol insecticide bombs may be used for bugs inside the house. Follow label directions, as always, when applying any of these insecticides.

## GRASSHOPPERS

*Phylum: Arthropoda      Class: Insecta      Order: Orthoptera*

Since most readers are already familiar with grasshoppers and can recognize them easily, these insects will not be discussed in great detail here. Suffice it to say that grasshoppers are very common and ubiquitous insects throughout the world. There are many types of grasshoppers, varying greatly in size, color, and habitat. Many grasshoppers are very destructive pests of agricultural crops, plants, shrubs, trees, and vegetables. These insects often swarm in vast hordes, devouring practically everything in their path.

For the average homeowner, grasshoppers usually pose a less serious problem. However, they often swarm in considerable numbers around houses during the warm months, and may devour the leaves of trees, bushes, shrubs, garden vegetables and other green plants.

### Control

Thoroughly treat all trees, shrubs, bushes, hedges, and other green plants and vegetables with a contact spray, residual spray, or dust. Smaller plants, garden vegetables, etc. also may be either sprayed or dusted for grasshopper control. Recommended insecticides are: 3% malathion, Sevin, Orthene, Lannate, or toxaphene; 1.5% Baygon, Dursban, diazinon, lindane, or Counter. NOTE: Check the pesticide container label for the exact recommended strength to apply.

### GRAY-BELLIED RATS (Roof Rats)

*Phylum: Chordata*    *Class: Mammalia*    *Order: Rodentia*

The gray-bellied rat, or roof rat, is the smaller of two common rat species occurring throughout the United States and many parts of the world. The roof rat is grayish-white to yellowish-white below. It measures six to eight and one half inches long, the tail is seven to ten inches, and the nose is pointed. This rat lives in the same areas as the larger Norway (ground) rat, but generally prefers urban areas and higher elevations. The roof rat is an adept climber, and frequents roofs, rafters, lofts, attics, trees, vines, wires, and other structures high above the ground.

Roof rats live nine to twelve months. They become sexually mature in three to five months, and may produce up to six litters of young per year. Generally, the roof rat has the same food requirements as the Norway rat, but usually eats higher percentages of vegetables and less meat and garbage.

### Control

For effective control of rats, it is essential to determine the type of rat involved. Control of roof rats involves different techniques and, in some cases, different poisons than those used for Norway rats. Another key point to emphasize is that rats will not linger where neither food nor water are available to them permanently. Thus, effective rat control involves general sanitation and preventive measures along with eradication techniques.

The following control measures apply to roof rats *only*:

1. If these rats infest your house or other buildings, attempt to locate the entrance point(s). These will be higher level (second floor or higher, attic, wires, cables, vines, limbs, etc.). Close off all suspected entrances that are practical to close.

2. Keep all garbage, including vegetables, in tightly-closed garbage cans.

3. Clean all spilled grain, nuts, vegetables, and other food material from floors and storage areas. Remove all trash, debris, and clutter.

4. Keep food in tightly-closed containers.

5. Eliminate all standing, or open, water containers.

The general sanitation measures listed above will go far toward controlling most rat infestations. In many cases, however, it will be necessary to use poison baits to completely control rats. Again, you must determine exactly which type of rat is involved: roof rats or the larger Norway rats.

The safest rodenticides are the commonly-used anticoagulant baits. These generally act slowly, however, and require several feedings for effective results. At this point, you must realize that rats are very wary and suspicious animals, and often pre-baiting is necessary for a period of time before poisons are actually applied. In some cases, several days are required before rats will eat a new source of food or bait. Recommended anti-coagulant poisons for roof rats are: warfarin, coumachlor, Fumarin, Pival, Pivalyn, PMP, diphacinone, chlorophacinone, Promar, and RoZol. If the rodenticide is not ready-mixed with the bait, use any of the following bait materials mixed with the poison: rolled oats, yellow corn-meal, bacon, salmon, mineral oil, salmon oil, or most human foods. The following liquid baits also may be applied as a water solution: Pivalyn, warfarin, Fumasol-G, or PMP.

Other than anti-coagulants, the following rodenticides are recommended: talon, zinc phosphide, and arsenic trioxide. NOTE: zinc phosphide and arsenic trioxide are extremely poisonous, and must be handled and applied with great care.

## GREEN JUNE BEETLES

See: *Flower Beetles* in this section.

## GROUND BEETLES

*Phylum: Arthropoda    Class: Insecta    Order: Coleoptera*

This is a very large group of common beetles with over 2500 species. They vary greatly in size, shape, and color—but most are darkish, shiny, and rather flattened in appearance. These beetles commonly occur under stones, logs, bricks, lumber, bark, and debris; or they may be seen running around on the ground or on the floors of houses and buildings. These beetles are fast runners.

Generally ground beetles are beneficial, feeding on other destructive insects such as caterpillars. Only in rare instances do these beetles become serious pests. Some species are called "bombardier beetles" because they eject from their anus a glandular fluid that vaporizes upon contact with the air and resembles a puff of smoke. This substance may irritate sensitive skin.

### Control

Rarely will control measures against ground beetles become necessary. However, when unusually large infestations occur in houses, yards, turfs, flowerbeds, etc., some insecticidal treatment may be necessary. A residual spray, contact spray, or dust is recommended, using any of the following insecticides: 1% lindane, Dursban, Baygon, or diazinon; or 2–3% malathion, Sevin, or methoxychlor.

## GROUND SQUIRRELS

*Phylum: Arthropoda    Class: Mammalia    Order: Rodentia*

Ground squirrels, often mistakenly called "gophers," may become pests in lawns, turfs, gardens, golf courses, and other areas. Active from March until September, they hibernate inside their burrows during the winter. Ground squirrels are nervous, excitable animals that resemble chipmunks, but are larger. They seldom venture far from their burrows which usually are located in open fields and brushy areas. Ground squirrels eat seeds, insects, roots, fruit, and green vegetation. In some cases, they may become troublesome pests, digging up newly planted seeds and causing damage to lawns and turfs.

### Control

Ground squirrels may be controlled in four basic ways: (1) Fumigation, (2) Repellents, (3) Traps, and (4) Poisons. Fumigation of burrows is the quickest and most humane way to control ground squirrels. Recommended fumigants are: liquid PCP, carbon tetrachloride, methyl bromide, liquid PDB (paradichlorobenzene), and carbon monoxide (automobile exhaust).

Recommended repellents are: PDB and naphthalene (mothballs or flakes).

Traps: rat snap traps, small animal traps, box traps, live traps, or Hav-a-Heart traps. Grain is the recommended bait for trapping these animals.

For appropriate poisons, see: *Chipmunks* (Control) and *Gophers* (Control) in this section.

## GRUBS

*Phylum: Arthropoda        Class: Insecta        Order: Coleoptera*

Grubs are the larvae of June beetles, chafers, and others. White grubs feed on the roots of grasses and plants, and are very destructive in some cases. They may do considerable damage to pastures, lawns, turfs, and crops such as: corn, small grains, potatoes, and strawberries.

### Control

Follow label directions on the container and apply any of the listed insecticides: aldicarb (Temik), Amaze, Aspon, bendiocarb, Furadan, chlormephos, Di-Syston, Mocap, Dasanit, Mesurol, Zectran, Thimet, or Counter.

## GULLS

See: *Blackbirds* (Control) in this section.

## GYPSY MOTH LARVAE

*Phylum: Arthropoda        Class: Insecta        Order: Lepidoptera*

The gypsy moth, first introduced into the United States from Europe about 1866, has become distributed widely throughout New England where it causes widespread damage to forest trees. The female gypsy moth is white with black markings and has a wingspan of about one-and-a-half to two inches. The male gypsy moth is slightly smaller. The female lays her eggs on tree trunks and similar spots. The eggs lie dormant during the winter and hatch the following spring. Female gypsy moths are weak fliers and rarely travel far from the cocoons from which they emerge. The young larvae (caterpillars) are chiefly responsible for the dispersal of this pest species.

### Control

See: *Caterpillars* (Control) in this section.

## HACKBERRY GALL PSYLLIDS

*Phylum: Arthropoda    Class: Insecta    Order: Homoptera*

Hackberry gall psyllids are tiny, flying, plant-eating insects. The nymphs (immature forms) infest hackberry trees in the spring and summer, often producing galls on the leaves and branches. The flying adults are attracted to lights, and can enter the house through ordinary fly screen on doors and windows. They may invade houses in great numbers.

### Control

Control should be directed primarily to the nymphs on hackberry and other trees in the spring and summer. However, control measures also may be necessary inside when adults invade the house in great numbers.

Treat infested trees for nymphal psyllids with a residual or contact spray. Also apply either of these spray forms to doors, screens, and windows. Recommended sprays are: 2% malathion or Sevin; 0.5% Dursban; or 1% diazinon or Baygon. For flying adults inside the house, use any of the following aerosol or space sprays: 0.5% DDVP; 0.15% SBP-1382®; 0.25% synergized pyrethrins (piperonyl butoxide); 1% Entex, Baytex, or Dibrom; or 1.5% trichlorfon or Gardona.

## HAM SKIPPERS

*Phylum: Arthropoda    Class: Insecta    Order: Diptera*

Ham skippers are the larvae of skipper flies. The adult flies are about one-fifth of an inch long, and are rather metallic black or bluish. The adults lay their eggs on cheese and preserved meat, and the larvae ("skippers") may become serious pests of these foods.

### Control

Control of skippers is best accomplished by fly-proofing food storage areas. Adult flies may be killed with an aerosol or space spray of: DDVP, pyrethrins, SBP-1382®, malathion, diazinon, Entex, or naled (Dibrom). Infested meat should be washed and rinsed thoroughly with clean water, and all storage rooms and facilities should be scrubbed and cleaned thoroughly. In some cases, it may be necessary to cover meats with plastic or cloth bags or wrap them with cloth, plastic, or paper.

## HARD TICKS

See: *Brown Dog Ticks* and *American Dog Ticks* in this section.

## HARVESTMEN

See: *Daddylonglegs* in this section.

## HARVEST MITES

*Phylum: Arthropoda      Class: Arachnida      Order: Acarina*

Adults are very common, harmless, red mites that swarm during the spring and summer in grass around homes, and in gardens, turfs, and other such areas. However, the nymphs of these mites are chiggers which are very irritating pests indeed.

### Control

When the adult red mites are visible, it may be advisable to treat for these in order to lessen the chigger population that will follow. Use the control methods, insecticides, and miticides recommended for *Chiggers* in this section.

## HARVESTER ANTS

*Phylum: Arthropoda     Class: Insecta     Order: Hymenoptera*

These are large, red to dark-brown ants, about one-fourth of an inch long, that gather seeds and grasses, hence the term "harvester ant." Although they do not actually invade houses, harvester ants nevertheless nest in areas surrounding houses, and may become serious pests. Nests are built with large craters in lawns, yards, turfs, gardens, etc. These are aggressive, pugnacious ants that inflict a very painful sting. The venom does not remain localized but spreads throughout the body lymph channels, often causing discomfort in the groin lymph glands.

### Control

Several insecticide formulations may be used to control harvester ants: sprays, dusts, granules, and baits. Sprays and dusts should be forced under pressure into nest craters, and residual sprays should be applied to lawns, turfs, gardens, flowerbeds, etc. where ants live. Baits and granules also may be applied in infested areas.

Recommended sprays and dusts are: 1% Knox Out, 0.25% Ficam, 0.5% Dursban or diazinon, 1.1% Baygon, or 1% malathion or Sevin.

Recommended baits and granules are: Baygon, Mirex, and Kepone.

## HEAD LICE

*Phylum: Arthropoda     Class: Insecta     Order: Anoplura*

Head lice are bloodsuckers, and are very similar to body lice. These pests infest humans throughout the United States and the world. They occur most abundantly on children, and spread rapidly through a family and throughout the community. Head lice eggs, called nits, are readily discovered when the hair and scalp are inspected closely. Nits commonly are attached to hair shafts close to the scalp behind the ears, but may occur on most any part of the scalp. Both the lice and the eggs must be destroyed for effective control.

### Control

Lice infestation of humans is often considered a medical problem and a physician can be consulted. However, you yourself can effectively treat and control both head lice and body lice infestations. The following procedure is recommended for head lice control:

1. Shampoo and dry the hair thoroughly at least once per day.

2. Seat the person being treated with the head tilted backward and the eyes covered with a towel or cloth.

3. Apply one of the insecticidal emulsions listed below. Apply this emulsion liberally, rubbing it well into the hair and scalp.

4. Comb the hair thoroughly.

5. If the pyrethrin emulsion has been used, wait fifteen minutes and shampoo the hair again. However, if lindane or benzyl benzoate was used, wait at least twenty-four hours before shampooing the hair again.

6. Following the second shampooing, dry, comb, and brush the hair thoroughly to remove all dead lice and loosened nits (eggs).

7. Repeat this procedure as needed.

Recommended insecticide emulsions for lice control are: 0.2% pyrethrins, 1% lindane (Kwell), or 12% benzyl bonzoate. NOTE: Children under twelve months of age should be treated by a physician.


## HIDE BEETLES

*Phylum: Arthropoda      Class: Insecta      Order: Coleoptera*

Adult hide beetles are black with yellow or whitish hairs scattered over the body. The shape of this beetle is very similar to the larder beetle, but slightly larger, averaging about one-third of an inch in length, with each wing apex tapering to a fine point. Larvae of hide beetles burrow into wood and building materials which they may damage structurally. Adult beetles infest hides, leather, skins, museum articles, and stored food in the kitchen and other areas.


### Control

Discard all infested food, cereals, meats, cheese, and other items. Remove all cooking utensils, dishes, and shelfpaper from all cabinets, cupboards, shelves, drawers, and pantries. Treat these areas with a residual spray or dust, paying close attention to all cracks and crevices and other hiding places.

Skins, hides, leather, etc. should be thoroughly cleansed with a suitable cleansing agent, and then treated with a very light coat of spray or dust.

The following insecticides are recommended for hide beetles: 1% lindane, Baygon, or diazinon; 0.5% Dursban; 0.25% to 0.5% Ficam; or 2% malathion or Sevin.

## HONEYBEES

See: *Bees* in this section.


## HORNETS

*Phylum: Arthropoda*      *Class: Insecta*      *Order: Hymenoptera*

Hornets belong to the group of insects called paper wasps. Hornets are medium- to large-sized insects, generally black, with whitish-yellowish markings on the face and body. They build a large gray-colored paper (cellulose) nest above the ground—usually in trees, bushes, or buildings. When they are disturbed, hornets will attack fiercely and sting viciously. These stings are very painful, and can be dangerous to sensitive (allergic) persons.


### Control

Ideally, hornets should be killed very quickly inside the nest before they can leave it and attack you. Preferably, this should be done at night when hornets are inactive. However, they are attracted to light and may attack you even at night if a light is used.

Several methods have been used to accomplish the quick destruction of hornets-and-nest. First, however, you are urged to protect yourself against these vicious insects. If possible, wear a beekeeper's suit that completely covers your body and leaves no skin exposed. If a beekeeper's suit is not available, then wear long, loosely-fitting trousers with the legs tucked inside a pair of boots, and a loosely-fitting shirt or jacket, gloves, and some type of plastic, screen, or cloth head mask that extends down inside the collar of your shirt.

1. Fashion a torch with a very hot, expanding flame at one end of a long pole. The flame should be as large and hot as possible. Scrap cloth, old clothes, discarded sacks, etc. can be wrapped around one end of the pole, secured there, then soaked with gasoline or kerosene, and lighted. NOTE: This method cannot be used in every instance because of the possible fire hazard it presents. In any case, you should have a garden hose or other water source nearby to control any fire that gets out of control. Applied correctly, this method will, in most cases, totally destroy a hornet nest and all insects inside it within a few seconds.

2. If you are well protected against stings, you can simply knock down the nest with a stick, allow the hornets to leave it, and then destroy it.

3. Insecticides sold in special squirting cans allow you to stand some distance away and squirt a fine stream of insecticide onto the nest and any emerging insects. Again, you should be well-protected from stings before attempting this method.

4. High-volume, high-pressure space sprayers may be used to thoroughly spray hornet nests and all emerging insects.

Recommended insecticides for quick knockdown are: pyrethrins, resmethrin, SBP-1382®, DDVP, Baytex, and Dibrom.

## HORNWORMS

*Phylum: Arthropoda*    *Class: Insecta*    *Order: Lepidoptera*

Hornworms are the larvae of sphinx or hawk moths. The adult moths are medium-sized to large, with heavy bodies and long, narrow front wings. In some cases, the wing span may extend five inches. The adult moth body is spindle-shaped, tapering, and pointed at both ends. The larvae (hornworms) get their name from the "horn" or spindle projecting from the top of the body. These larvae pupate in the ground, and may do extensive damage to plants and vegetables, especially tomatoes and tobacco.

### Control

See: *Caterpillars* (Control) in this section.

## HORSEFLIES

*Phylum: Arthropoda*    *Class: Insecta*    *Order: Diptera*

Horseflies are very similar to deerflies, except larger. Both types are very annoying pests of man and animals. Many horsefly species are vicious biters, and inflict painful bites that itch for several days. This group of flies (tabanids) can serve as vectors (transmitters) of various diseases to man and animals, such as Q-fever, anaplasmosis, filariasis, hog cholera, infectious anemia, California encephalitis, anthrax, and tularemia.

### Control

See: *Deerflies* (Control) in this section.

## HOUSEDUST MITES

*Phylum: Arthropoda    Class: Insecta    Order: Acarina*

These mites tend to occur in damp, moist places frequented by people and pets. They occur in houses, hotels, motels, hospitals, retail stores, and other such areas. These mites live inside mattresses, pillows, upholstered furniture, and similar places. They also live in rugs, carpets, and cracks and crevices in floors. These mites are very small, and cannot be seen with the naked eye. They produce an allergic, asthma-like reaction in some sensitive (allergic) persons.

### Control

No chemical treatment has been developed to control housedust mites. However, vacuuming and electronic dust filters may provide some relief.

## HOUSEFLIES

*Phylum: Arthropoda    Class: Insecta    Order: Diptera*

Houseflies are among the most widespread, abundant, and persistent of all pests that affect mankind. The female housefly begins laying eggs about four to twenty days after emergence as an adult. These eggs are small, white, and oval-shaped, and are laid in batches of seventy-five to about one hundred fifty each. They hatch in twelve to twenty hours during warm conditions. Fly larvae, called maggots, burrow into any surrounding filth, including human and animal feces, dead animals, spoiled meat and other food, decaying vegetation, and other organic matter. Under normal conditions, the entire housefly life cycle is completed in eight to twenty days.

Houseflies are extremely filthy because of their living habits, and can accidentally transmit disease organisms from septic, filthy areas to humans. For example, the organisms that cause typhoid, dysentery, cholera, and other gastro-intestinal disturbances can be transmitted by houseflies in some cases.

### Control

Good sanitation is basic to effective housefly control. Eliminate all fly-breeding sources, such as animal manure, pet feces, open garbage dumps or cans, dead animals, and other types of exposed organic matter. Fly-proof all doors and windows of houses and buildings that you wish protected from houseflies.

For flying adults inside houses and buildings, use an aerosol or space spray of: allethrin, pyrethrins, dichlorvos, or resmethrin. If the infestation is unusually heavy, or continues, apply a residual spray to walls, floors, ceilings, and other surfaces frequented by flies. The following insecticides are recommended: 1% dimethoate; 3% malathion; 1% diazinon; 1% ronnel; 2% Rabon; 1% pyrethrins, resmethrin, or allethrin.

The following wet and/or dry baits also are recommended for fly control outdoors: 0.1% dichlorvos; 0.2% (wet) or 2% (dry) malathion; or 2% (dry) dimethoate, trichlorfon, or ronnel.

For larval (maggot) control, apply a residual or contact spray of 1–2% dimethoate, diazinon, dichlorvos, or malathion.

## HOUSEPLANT PESTS

In most cases houseplant pests are mites, spider mites, springtails, thrips, aphids, scale insects, whiteflies, gnats, and ants.

### Control

Wash all plants before bringing them inside. This often eliminates the need for insecticidal treatment later. However, heavy or persistent infestation of plants by pests may require occasional chemical treatment. Generally the following sprays or dusts will be effective against houseplant pests: 1–2% dicofol, lindane, malathion, rotenone, diazinon, Gardona, Aramite, Plictran, Mitox, and Kelthane.

## HUMAN FLEAS

See: *Fleas* in this section.

## IMPORTED FIRE ANTS

See: *Fire Ants* in this section.

## INCHWORMS

See: *Caterpillars* in this section.

## INDIAN MEAL MOTHS

*Phylum: Arthropoda*     *Class: Insecta*     *Order: Lepidoptera*

The Indian meal moth is a pest of stored food inside the house where it feeds on grains, flour, cornmeal, dried fruits, spices, and other food material. This moth may be mistaken for the clothes moth, but distinctive markings on the meal moth wings will enable you to distinguish the two. The Indian meal moth forewings are tannish-brown in the front section, reddish-brown on the back section, and the span is about three-fourths of an inch. The Indian meal moth tends to rest in dark, secluded places during the day, and when disturbed, flies in an irregular, zig-zag manner. Larvae of this moth are wormlike, about one half of an inch long and are dirty white to greenish or pinkish in color.

### Control

Discard all infested food material. Remove all dishes, cookware, food containers, and shelfpaper from all cabinets, cupboards, pantries, shelves, and drawers. Treat these areas with a residual spray or dust of: 1.1% Baygon, 0.5% Dursban or diazinon, 0.25% Ficam, or 2% malathion.

Adult moths should be treated with a contact or space spray of: 1% SBP-1382®, DDVP, resmethrin, or pyrethrins.

## JAPANESE BEETLES

*Phylum: Arthropoda*     *Class: Insecta*     *Order: Coleoptera*

The Japanese beetle is a serious pest of lawns, golf courses, fruits, and shrubbery, particularly in the eastern United States. These beetles are brightly colored and very attractive, with a bright metallic-green head and thorax, a brownish abdomen tinged with green on the edges and with spots along the sides. Larvae of Japanese beetles live in the soil where they feed on plant roots, while the adults feed on the leaves, foliage, and fruit of plants. Both larvae and adults may become destructive pests of plants.

**Control**

For adult beetles on the foliage of plants, apply a residual spray or dust of: 1–2% lindane, Sevin, or malathion; or 1% Ficam, Dursban, Baygon, or diazinon.

For larvae in the soil (rootworms) follow label directions and apply any of the following insecticides: Furadan, Amaze, ethoprop, Dasanit, fonofos, Thimet, or Counter.

## JUNE BEETLES (June Bugs or May Beetles)

*Phylum: Arthropoda     Class: Insecta     Order: Coleoptera*

This is a large group of plant-eating beetles that are serious pests of plants, flowers, and cultivated crops. June beetles and/or May beetles, often called "June bugs," generally are brown or green in color, and swarm around lights in spring and early summer. The larvae of these beetles are white grubs which feed on the roots of grasses and plants within the soil. Both adults and larvae may become serious and destructive pests of flowers, plants, and other vegetation.

**Control**

Control of these beetles should be directed at both the adults which feed at night on plant foliage, and at the larvae (grubs) within the soil. For adult control, treat infested plants with a residual spray or dust of: 1.5% Baygon; 1–2% lindane; 2% malathion, diazinon, or Sevin; or 1% Dursban.

For larval control, see: *Grubs* in this section.

Repeat all treatments as needed.

## KISSING BUGS

See: *Assassin Bugs* in this section.

## LADYBIRD BEETLES (Ladybug Beetles or Lady Bugs)

*Phylum: Arthropoda     Class: Insecta     Order: Coleoptera*

Generally this group of harmless, brightly-colored, and attractively marked beetles are beneficial because they feed on aphids, plant lice, and other harmful insects on flowers, vegetables, and other plants. At times, however, ladybird (ladybug) beetles do become pests when they invade houses in large numbers.

### Control

Control measures should not be applied against ladybird beetles unless they swarm in large numbers and/or invade houses and buildings. If this occurs, however, remove all leaves and debris from around houses and building foundations, and from around windows. Large infestations of beetles on the ground, or on the walls of buildings, may be controlled with a residual or contact spray of: 1–2% lindane, malathion, diazinon, Sevin, Baygon, or Dursban. Beetles inside the house may be controlled with the same chemicals at a 1% strength. Aerosol bombs specifically listed for beetles also may be used indoors.

## LARDER BEETLES

*Phylum: Arthropoda*　　*Class: Insecta*　　*Order: Coleoptera*

The larder beetle is similar to the hide beetle, and both pests may live in the same general areas. The larder beetle is a common pest of cured meats in Europe, Canada, and the United States. The adult beetle is about one-third of an inch long and dark brown, with a pale-yellow six-spotted band on the back and underbelly. Larvae are about one half of an inch long and brownish, with two curved spines on the last segment of the body. These pests attack ham, bacon, cheese, various other meats, museum specimens, stored tobacco, dried fish, seafood, and even dog biscuits. Adults tend to enter the house in May and June, and search for stored food to eat where they also lay their eggs. Cracks, crevices, and other similar hiding places in and around cupboards and pantries are the preferred quarters of these beetles.

### Control

Discard all infested food. Remove all pans, dishes, cookware, stored food, and shelf paper from all cabinets, cupboards, pantries, shelves, and drawers. Treat these areas with a residual spray of: 2% malathion, 0.25% Ficam, 1.1% Baygon, 0.5% Dursban or diazinon; or a contact spray of: DDVP, pyrethrins, or resmethrin.

## LATRINE FLIES

*Phylum: Arthropoda     Class: Insecta     Order: Diptera*

Latrine flies are very similar to sewer flies and drain flies. This fly is smaller than the housefly, and has a bluish-black thorax. These flies usually appear in the early spring before houseflies become numerous. The latrine fly prefers human feces as a food source and breeding habitat, but also may live in decaying vegetation and other exposed animal dung and organic matter. These flies may become pests in houses and buildings.

### Control

Use the control procedures and insecticides recommended for *Blowflies* and/or *Houseflies* in this section.

## LEAF BEETLES

*Phylum: Arthropoda     Class: Insecta     Order: Coleoptera*

These beetles generally are one half of an inch (or less) long, and many are brightly colored. Adults feed on the foliage of plants and flowers. The larvae also are plant-eaters, but may vary in their feeding preferences. Some attack plant foliage, some feed on the roots, and others bore into the plant stems. These beetles are serious pests of flowers and cultivated plants. Most species spend the winter in the adult form.

### Control

Since these beetles mostly overwinter as adults, you should attempt to eliminate them before winter begins, thus possibly preventing, or lessening the severity of, a new generation of beetles the following year. Treat infested plants and flowers until the foliage turns brown in the fall, using a residual spray or dust of: 1–2% lindane, Baygon, Dursban, diazinon, malathion, or Sevin.

## LEAF BLOTCH MINERS

*Phylum: Arthropoda     Class: Insecta     Order: Lepidoptera*

Leaf blotch miners are the larvae of a large group of small to minute moths. These larvae (leaf miners) typically make blotch mines in the leaves of plants, causing the leaves to fold or roll up. The white oak leaf miner is common in the eastern United States, and feeds on various kinds of oaks. Leaf miners generally live on the upper leaf surfaces.

### Control

Since these larvae overwinter in dry leaves, it is advisable to burn or dispose of all dead, dry leaves around the house in the fall to prevent, or lessen the severity of, a new leaf miner crop the following year.

Treat affected plants with any of the listed insecticides: 2–3% diazinon, trichlorfon, Orthene, Pounce, Ambush, Pydrin, or Sevin; or apply according to label directions: dimethoate, demeton, or Meta-Systox-R.

## LEAF CHAFERS

See: *June Beetles* and *Japanese Beetles* in this section.

## LEAF-CUTTING ANTS

*Phylum: Arthropoda       Class: Insecta       Order: Hymenoptera*

Leaf-cutting ants are known as "parasol" ants because they characteristically carry bits of leaves over their heads umbrella-fashion. They are very destructive pests of trees and plants, and have been known to defoliate an entire tree overnight.

Leaf-cutting ants vary in size, from large to very small, and tend to live in areas from Louisiana and Texas, southward through Mexico, into South America. The large workers among leaf-cutting ants are fierce biters.

### Control

Control measures against these ants should be directed at both their nests and the affected plants, trees, shrubs, etc. on which they feed. For treating ant nests, use a pin-stream nozzle and force a large amount of residual spray into the nest openings. A dust or powder may be substituted by washing it into the nest openings with a hose. If nests cannot be located, or happen to be inaccessible, treat the ground around the nests and around plants, trees, shrubs, and other vegetation with a strong residual spray or dust. Treat affected plants, trees, and vegetation with the same residual spray.

Recommended insecticides are: 1–2% diazinon, Baygon, or Dursban; or 2–3% malathion, Sevin, methoxychlor, or Gardona.


# LEAFHOPPERS

*Phylum: Arthropoda     Class: Insecta     Order: Homoptera*

Leafhoppers are a very large group of plant-eating insects that are serious pests of many cultivated plants, flowers, shrubs, trees, grasses, and field and garden crops. Leafhoppers vary in size from a fraction of an inch to about one half of an inch in length, and many have beautiful colored markings. These insects feed primarily on the leaves of host plants, causing five (5) major types of injury. Some remove excessive amounts of sap and juices, causing the leaves to appear spotted and eventually turn yellowish or brownish. This type of damage occurs primarily on apple trees. Another type of injury produced by leafhoppers is browning of the outer portion of leaves, eventually causing the entire leaf to turn brown. Some leafhoppers cause the terminal part of twigs to die, while others are vectors (transmitters) of plant diseases such as aster yellows, "curly top" in beets, yellow dwarf in potatoes, necrosis of elm , Pierce's grape disease, and corn stunt.

## Control

Apply to affected trees, plants, shrubs, vegetables, etc. a residual or contact spray, or a dust, using any of the listed insecticides: Sevin, methoxychlor, acephate, Meta-Systox-R, Systox, Di-Syston, or Dimecron. Follow label directions.


# LEAF MINERS

*Phylum: Arthropoda     Class: Insecta     Order: Lepidoptera and
                                                    Diptera*

Leaf miners are the larvae of various flies and moths that exist in many parts of the world. The adult flies are small and usually blackish to yellowish in color. Eggs are laid on the leaves of plants, trees, and other vegetation. After hatching, the larvae form serpentine mines in the leaves. See also: *Leaf Blotch Miners* in this section.

### Control

Follow label directions and apply to affected plants, trees, and other vegetation any of the listed insecticides: diazinon, trichlorfon, malathion, dimethoate, Sevin, demeton, Meta-Systox-R, or acephate.

## LEAF ROLLERS

*Phylum: Arthropoda     Class: Insecta     Order: Lepidoptera*

Leaf rollers are the larvae of tortricid moths. Adult moths are usually gray, tan, or brown, with a spotty, mottled appearance. A common pest in this group is the fruit-tree leaf roller, which makes a leaf nest in fruit and forest trees, and may cause serious defoliation of some trees. Another pest species in this group is the spruce budworm which attacks, and seriously damages spruce, fir, balsam, and other evergreens. These pests may completely defoliate and/or kill a large number of trees.

### Control

See: *Caterpillars* (Control) in this section.

## LEATHER BEETLES

See: *Hide Beetles* and *Larder Beetles* in this section.

## LESSER GRAIN BORERS

*Phylum: Arthropoda     Class: Insecta     Order: Coleoptera*

This beetle is an important pest of stored grain in the United States, Canada, Argentina, India, Australia, and other areas of the world. It also infests rice, wood, and books. The adult beetle is about one-eighth of an inch long, dark brown to black in color, with tiny pits on the back.

### Control

Large quantities of infested grain or wood, or large numbers of books will require fumigation for effective results. Large-scale fumigation must be done by a licensed pest control operator.

Smaller infestations of lesser grain borers inside the house, however, may be treated by the average person. Remove all pots, cookware, dishes, stored food, grain, cereals, and shelf paper from all infested cabinets, cupboards, pantries, shelves, and drawers. Remove infested books from shelves. Treat these areas with a residual spray or dust, paying close attention to all cracks, crevices, and other hiding places. The following insecticides are recommended: 1–2% DDVP, pyrethrins, or resmethrin for visible beetles; 2% malathion, 1.1% Baygon, 0.25% Ficam, or 0.5% Dursban or diazinon as a residual treatment.

## LICE, HUMAN

See: *Body Lice, Crab Lice*, and *Head Lice* in this section.

## LIGHT-ATTRACTED INSECTS

See: *Alderflies, Dobsonflies, June Beetles, Fishflies, Moths, Lovebugs,* and *Flying Insects* in this section.

## LOCUSTS

*Phylum: Arthropoda      Class: Insecta      Order: Orthoptera*

True locusts (*not* cicadas) are very similar, and closely related to, grasshoppers. Throughout recorded history, locusts have caused widespread crop losses, famine, plague, and pestilence when they have moved over the land in colossal hordes, devouring all green vegetation in their path. For the average person, however, locusts present a much less dramatic problem. For all practical purposes, locusts should be treated the same as grasshoppers.

### Control

Use the control measures recommeded for *Grasshoppers* in this section.

## LONE STAR TICKS

*Phylum: Arthropoda*     *Class: Arachnida*     *Order: Acarina*

The lone star tick belongs to the group known as hard ticks. Female lone star specimens are recognized easily by a silvery-white spot at the tip of the shield (scutum) on the back, hence the term "lone star" tick. This tick bites man readily in the larval, nymphal, and adult stages. The bite is painful and may itch for several days. The lone star tick is a vector of Rocky Mountain spotted fever, tularemia, and perhaps Q fever as well. It may exist on livestock, dogs, deer, birds, and humans.

### Control

See: *Brown Dog Ticks* (Control) and *American Dog Ticks* (Control) in this section.

## LOOPERS

*Phylum: Arthropoda*     *Class: Insecta*     *Order: Lepidoptera*

Loopers are the larvae of noctuid moths. These larvae (loopers) are smooth and dull-colored, with three pairs of short leg-like projections on the body. They attack and damage cabbage, vegetables, and other plants. Loopers resemble inchworms in their movement.

### Control

See: *Caterpillars* (Control) in this section.

## LOVEBUGS

*Phylum: Arthropoda*     *Class: Insecta*     *Order: Diptera*

Adult lovebugs are black, with black wings, while the top of the thorax is red. They are slightly over one half of an inch long. These flies have various names such as March flies, honeymoon flies, telephone bugs, and doubleheaded bugs. They live throughout the Gulf States, Mexico, and Central America. Although lovebugs do not bite or sting humans, they nevertheless swarm in great numbers, often creating a nuisance when driving at night. These insects splatter automobile headlights, paint, and windshields, and even clog up radiators, causing engine overheating and damage in some instances. They also swarm around patios and other outdoor lighted areas during the warm months.

### Control

Flying adults around houses, buildings, patios, pools, etc. may be partially controlled by using the aerosol, space, or residual spray form of insecticides listed for *Flying Insects* in this section. Electronic bug killers installed around patios, pools, and other infested areas also may prove helpful in reducing the number of lovebugs, as well as other flying insect pests.

# MAGGOTS

*Phylum: Arthropoda*    *Class: Insecta*    *Order: Diptera*

Maggots are the larvae of flies. For the average person, housefly larvae (maggots), in particular, present the most serious infestation. Maggots tend to remain localized where they hatch, however, feeding on any filth that surrounds them, like human and animal feces, horse and cow manure, exposed garbage, sewers, dead animal carcasses, and other types of organic material. Occasionally, however, maggots do present a direct infestation problem in and around the house. Mature maggots, just before pupation, are about one half of an inch long, conical shaped, with dark mouth hooks, and creamy white in color.

### Control

Control of maggots should begin with control of the adult flies which produce them. Both maggots and flies can be reduced greatly by practicing good sanitation. This includes keeping all pet feces, animal feces, horse and cow dung, and dead animal carcasses cleaned up. It also means keeping all garbage in tightly-closed containers, and removing all decaying vegetation and other organic matter.

Individual maggot infestations that occur occasionally should be treated with a residual or a control spray of: 2% diazinon, malathion, dimethoate, Cygon, dichlorvos, fenthion, naled, DDVP, ronnel, or Gardona.

## MANTIDS

*Phylum: Arthropoda*    *Class: Insecta*    *Order: Orthoptera*

Mantids are large, elongated insects that are related to grasshoppers and walking sticks. One common species of mantid is the praying mantis. Generally, mantids are beneficial because they feed on other harmful insects. Occasionally, however, they may become pests when they appear in large numbers in yards, turfs, and on patios and porches, etc. Mantids are harmless to humans and animals.

### Control

Very rarely should chemical control of mantids become necessary. When occasional control is needed, however, use the procedures and insecticides recommended for *Grasshoppers* in this section.

## MARCH FLIES

See: *Lovebugs* in this section.

## MEAL MOTHS

See: *Indian Meal Moths* in this section.

## MEALWORMS (Yellow Mealworms/Dark Mealworms)

*Phylum: Arthropoda*    *Class: Insecta*    *Order: Coleoptera*

The yellow mealworm and the dark mealworm are the two most common forms of these pests. Both are the larvae of beetles, and are very similar, except for color. These two mealworms are nocturnal (active at night) and seek dark, moist, undisturbed places in which to hide. Mealworms live in warehouses, grain and meal storage bins, feedstores, in the litter of chickenhouses and birdhouses, and occasionally in kitchen pantries. The adult beetles are shiny, vary from dark brown to black, and are about one half of an inch long.

Both the mealworms and the eggs from which they hatch may be ingested in contaminated food, particularly cereals and breakfast foods. This condition, called *canthariasis*, often produces gastrointestinal disturbances.

### Control

Discard all contaminated food. Clean up all spilled grain, cereals, meal, etc. Clean up all damp, moist areas and vent them. Stack bags of grain, cereals, meal, etc. on racks above the floor to allow air circulation and drying.

For kitchen infestations, discard all contaminated foods. Remove all dishes, pans, cookware, foods, and shelf paper from all pantries, cabinets, cupboards, shelves, and drawers. Treat these areas with a residual spray or dust, paying close attention to all cracks, crevices, and other hiding places. The following insecticides are recommended: 1.1% Baygon, 0.5% Dursban or diazinon, 0.25% Ficam, 2% malathion or Sevin, or 1% lindane.

## MEALYBUGS

*Phylum: Arthropoda      Class: Insecta      Order: Pseudococcidae*

Mealybugs are soft-bodied insects with a white, powdery *wax* covering. They are about one-fourth of an inch long, and tend to live along the veins on the underside of leaves of plants. In addition to sucking plant juices, mealybugs also cause the "sooty-mold" fungus that discolors many plants. Continuous feeding by these pests may stunt the growth of plants and cause them to wilt.

### Control

Mealybugs can be rubbed off plants by hand, or washed off with a hose. They also may be removed from plants by swabbing the leaves with an alcohol-soaked sponge. Alternatively, the following chemicals are recommended in contact or residual spray form: 1–2% malathion, Sevin, or diazinon; or follow label directions and apply: Temik, Bidrin, Meta-Systox-R, Di-Syston, or Trithion.

## MEASURING WORMS

See: *Caterpillars* in this section.

## MELON WORMS

*Phylum: Arthropoda*     *Class: Insecta*     *Order: Lepidoptera*

Melon worms are the larvae of moths. Adult moths are rather large and have conspicuously-marked wings of glistening white with a black border. The larvae (melon worms) feed on the foliage, and burrow into the stems, of various melons and other plants.

### Control

See: *Caterpillars* (Control) in this section.

## MEXICAN BEAN BEETLES

*Phylum: Arthropoda*     *Class: Insecta*     *Order: Coleoptera*

The Mexican bean beetle is a large ladybird beetle that is yellowish in color, with eight spots on each side of the back. Both the adults and larvae are plant-eaters, and are destructive pests of garden vegetables and other plants.

### Control

Treat affected plants with a residual spray, or dust, of: 2–3% rotenone, malathion, Sevin, diazinon, ryania, sabadilla, Gardona, or lindane. Repeat this treatment as needed.

## MICE

*Phylum: Chordata*     *Class: Mammalia*     *Order: Rodentia*

Because most readers are already familiar with mice, these pests will be discussed only briefly here.

Although mice consume much less food than rats, they nevertheless are serious pests because of their constant gnawing, nibbling, and contamination of food materials, and because of the damage they do to woodwork and other household items. Moreover, a large infestation of mice produces a disagreeable, musty order in houses and other buildings. Mice become sexually mature between two and three months of age, and they usually live two to three years. Litters vary from about five to perhaps fifteen young. Although mice do not actively seek water, they will drink small amounts when it is available. Moreover, when mixed with poison bait, water increases its attractiveness to mice.


### Control

As with rats, effective mouse control begins with good sanitation. It is suggested that you follow the general sanitation recommendations for controlling *Gray-Bellied (Roof) Rats* given in this section.

Mice are more likely to become household pests than are rats. Mice inhabit closets and bedding, and especially the insulation between walls and in attics, which makes ideal nests. Following application of the recommended sanitation practices covered earlier, a mouse infestation that persists will require the use of poison bait and/or trappings.

Recommended poisons for mice are: warfarin (D-Con and other name-brands), talon, zinc phosphide, arsenic trioxide, chlorophacinone, diphacinone, PID®, fumarol, Fumasol, Pival, Pivalyn, and PMP. Baits may be either dry or wet. Remember, however, that water mixed with poison baits increases their attractiveness to mice.

Some poisons will contain bait already mixed with them, while others will require that you mix an acceptable bait with the poison before applying it. Some recommended mouse baits are crushed grain, cornmeal, and seeds.

Poison baits should be placed in numerous stations throughout the house because mice do not travel very far from their nests. Several bait stations should be placed in kitchens, cabinets, closets, basements, attics, and other areas where there is evidence of mice.

Glue boards and mouse traps may be effective in controlling mice when the infestation is limited.


## MIDGES, BITING

*Phylum: Arthropoda     Class: Insecta     Order: Diptera*

Biting midges are very tiny flies, only a fraction of an inch long. Often called "punkies" and/or "no-see-ums," many of these insects bite man viciously and suck blood. They can be very annoying pests. Because they breed in large and extensive areas of water, midges unfortunately are difficult to control effectively.

**Control**

Follow the control procedures recommended for *Gnats* and *Black Flies* in this section. These measures are given to provide temporary relief only.

## MILDEW

Mildew results from the presence of tiny parasitic fungi, which are living plants. Mildew attacks various woods, cotton, linen, paper, leather, and other things that are exposed to a damp, moist environment. Mildew causes unsightly stain and discoloration, and it may ruin some things that it attacks. Mildew does not damage wood.

**Control**

As with wood-decaying fungi and other fungal forms, control of mildew begins with proper air ventilation, water drainage, and drying of all infested areas and articles. Mildew stains may be alleviated, or lessened, with a solution containing approximately one part bleach (chlorine, sodium hypochlorite, etc.) in two parts of water.

## MILLIPEDES

*Phylum: Arthropoda     Class: Diplopoda*
*Orders: Polyxenida, Glomerida, Polydesmida, Platydesmida,*
*Polyzoniida, Chordeumida, Julida, Spirobolida, Cambalida,*
*Spirostreptida*

Millipedes are long, cylindrical, worm-like arthropods with many body segments and many pairs of legs, hence the nickname "thousand leggers." Adults may vary from about one half of an inch up to three or four inches in length. Millipedes also vary in color, from blackish and brownish to red, orange, or even mottled. Millipedes move slowly, contrasting greatly with fast-moving centipedes. Like sowbugs and pillbugs, millipedes require a high-moisture environment, and all three of these pests may exist together, particularly in flowerbeds, mulch beds, moist gardens, etc., where they feed on decaying vegetation, insects, and earthworms.

Occasionally millipedes become pests when they invade houses in large numbers, especially following excessive drought or excessive rainfall. Invasion of houses by millipedes also may occur in late fall when they seek a sheltered area in which to overwinter. Some millipedes secrete a blister venom that may cause injury to the skin and, especially, to the eyes should it be squirted into them.

### Control

Inside, use a residual or contact spray of: 1.1% Baygon, 0.5% Dursban or diazinon, 0.25% Ficam, or 2% Sevin or malathion. Dusts, however, are better for wall voids than are sprays.

Outside, remove decaying leaves, plants, mulch, and other organic matter. Treat flowerbeds, mulch areas, damp soil, and other such areas with: Mesurol, Zectran, or metaldehyde, following label directions. Also, the insecticides listed for inside usage may be applied outdoors as well. Finally, apply a residual spray barrier five to fifteen feet wide around the house or building perimeter. Also treat the foundation, lower walls, crawl spaces, steps, and porches.

## MITES

*Phylum: Arthropoda     Class: Arachnida     Order: Acarina*

Mites are very small arachnids (not insects) with two body parts, four pairs of legs, and sucking mouthparts. The mite life-cycle consists of egg, larva, nymph, and adult. Nymphal mites, however, have only six legs. The life-cycle is usually completed in two to three weeks. Under favorable conditions, tremendous numbers of mites are produced in a relatively short time. Mites are parasites of many types of animals, birds, and plants. They are destructive pests of cultivated plants, crops, foods, trees, fruit, and other vegetation.

### Control

A large number of miticides and insecticides are effective against mites. Always follow label directions when treating for mites indoors and outdoors, whether on plants, trees, shrubbery, fruits, vegetables, lawns, buildings, or animals. The following pesticides are recommended: Mitox, Aramite, Kelthane, Thiodan, fenson, Trithion, dioxathion, EPN, ethion, Omite, Ovex, Morestan, Plictran, Tedion, Vendex, Pectac, Galecron,

Fundal, malathion, diazinon, Dursban, Baygon, Acarben, pyrethrins, and resmethrin.

## MOLDS

Molds are fungi which attack wood and other materials under damp, moist conditions. While mold is not harmful to wood, it does cause staining and discoloration. However, the presence of mold indicates that conditions are right for the more serious wood-decaying fungi to thrive.

### Control

As with most other fungi, mold can be controlled by eliminating dampness and moisture, improving water drainage, and increasing air circulation.

## MOLES

*Phylum: Chordata*     *Class: Mammalia*     *Order: Insectivora*

Moles are small subterranean mammals that are nearly blind, and rarely appear above ground. They burrow deep into the ground, and often become pests when their surface tunneling defaces lawns, gardens, turfs, golf courses, and other such areas. Moles also damage ornamental plants, shrubs, bushes, etc., by destroying the roots with their extensive tunneling.

### Control

Moles are controlled best by fumigating their underground tunnel network. Fumigation, however, must be done by a licensed pest control operator in most cases.

Small-scale, limited-area fumigation for moles may be carried out by the average person. First, remove the top dirt from a tunnel until the main passageway is uncovered. If a liquid fumigant is to be used, pour it liberally into the tunnels at ten- to fifteen-foot intervals until all the burrows are treated. Re-cover each treatment hole, but do not pack the dirt tightly enough to seal off the tunnel.

If a gas fumigant, such as carbon monoxide, is to be used, uncover the main passageway of each burrow and insert the free end of the hose leading from your automobile exhaust pipe, place dirt around the hose to prevent gas from escaping, and allow the gas to be pumped under pressure into the tunnel system for approximately one-half hour.

The following chemicals are recommended to fumigate moles: liquid PDB (paradichlorobenzene), methyl bromide, carbon tetrachloride, and carbon monoxide (automobile exhaust gas). NOTE: More than one treatment may be required with any of these chemicals.

# MOSQUITOES

*Phylum: Arthropoda*    *Class: Insecta*    *Order: Diptera*

Certainly all readers are familiar with mosquitoes, one of mankind's perennial, most annoying, and most dangerous insect pests. Mosquitoes are true flies, with only one pair of wings, and there are more than 2500 different species. Moreover, they are the most important arthropod vectors of diseases that affect humans and, thus, present a direct threat to man's health, not to mention his comfort and peace of mind.

Some diseases transmitted to man by mosquitoes are Saint Louis encephalitis, western equine encephalitis, eastern equine encephalitis, California encephalitis, Venezuelan equine encephalitis, dengue, yellow fever, malaria, and filariasis. Because of the ever-present danger of epidemics spread by mosquitoes, millions of dollars are spent yearly by local health departments and mosquito control districts to control mosquitoes and, thus, to safeguard the health and comfort of people throughout the world. It is estimated that mosquitoes have, throughout history, caused the death of more people than have all wars of mankind combined.

Mosquitoes undergo complete metamorphosis: egg, larva, pupa, and adult. The first three stages always occur in water. The adult (fourth stage) is a flying insect that feeds upon the blood of man and animals, and upon various plant juices. Only the female mosquito bites man and animals because blood is required to develop her eggs. Mosquito eggs generally are laid in rafts on the surface of water, but some species lay individual eggs. Larvae, which undergo development through several instars, are called "wigglers."

Hordes of mosquitoes breed in swamps, drainage ditches, sewers, storm drains, lagoons, and other such areas with standing water. Large numbers of mosquitoes also are produced around houses in standing water containers, such as rain barrels, clogged drainage gutters, tin cans, fish bowls, fish ponds, discarded tires, and pot holes.

### Control

While mosquito control generally is the province of community and local governmental agencies, individuals can contribute significantly to the reduction of mosquito breeding around their homes. During the spring, summer, and fall months, keep all outside water containers emptied, or treat the water surface with a recommended mosquito larvicide.

Flying mosquitoes inside may be treated with any of the insecticides recommended for *Flying Insects* in this section.

Recommended *larvicides* for treating standing water are: Abate, Altosid, diesel fuel, fuel oil, kerosene, and crankcase oil.

## MOTHS

See: *Casemaking Moths, Webbing Clothes Moths, Clothes Moths, Carpet Moths, Indian Meal Moths, Armyworms, Cutworms, Cankerworms, Leaf Rollers, Melon Worms,* etc. in this section.

## MOTH FLIES

See: *Drain Flies* in this section.

## MUD DAUBERS

*Phylum: Arthropoda      Class: Insecta      Order: Hymenoptera*

Mud (dirt) daubers belong to the group of insects known as thread-waisted wasps. These insects are very common. Most species are one inch or more long, and build nests of mud on the walls and ceilings of buildings. Mud daubers generally do not become serious pests. Rather, they are actually beneficial because they eat spiders.

### Control

Control measures against mud daubers rarely, if ever, should become necessary. Generally they do not congregate in large enough numbers to become pests. However, in rare cases when they build large numbers of unsightly mud nests on walls and ceilings, simply use a stick to knock down the nests. If this action fails to correct the problem, then apply a residual spray to the affected surface areas.

Recommended sprays are: 5–10% Sevin or malathion; 1.5% Baygon; or 1% fenthion, Dursban, or diazinon.

When old or unpainted buildings are involved, a residual coat of diesel fuel or crankcase oil may be effective in repelling mud daubers.

## MUSEUM BEETLES

See: *Hide Beetles* and *Larder Beetles* in this section.

## NEMATODES

*Phylum: Aschelminthes*    *Class: Nematoda*    *Orders:* Various

Nematodes are not arthropods and, therefore, they are neither insects nor arachnids; rather, nematodes are true worms. Specifically, they are roundworms that are serious parasites of many animals and plants. An example is the sugarbeet nematode which attacks sugarbeets, often doing serious damage or even destroying these plants.

Generally nematodes deposit their eggs either in the roots of the host plant or in the soil. When the eggs hatch, the worms occur within the plant cells and roots. They devour the internal tissue, often creating galls or "root knots." In some cases the plants die.

### Control

For plant nematodes, follow label directions and apply any of the listed nematocides: aldicarb, carbofuran, EDB, phenphene, Nemacide, Di-Syston, Mocap, Dasanit, Thimet, Nemofos, Vapam (SMDC), or MIT (Vorlex).

## NIGHT-FLYING INSECTS

See: *Alderflies, Moths, Dobsonflies, Lovebugs, Mosquitoes,* and *Light-Attracted Insects* in this section.

# NITS

See: *Head Lice* in this section.

# NO-SEE-UMS

See: *Midges* (*Biting*) in this section.

# OAK GALLS

*Phylum: Arthropoda*     *Class: Insecta*     *Order: Hymenoptera*

Oak galls, as well as many other types of plant and tree galls, are produced by very small (usually black) wasps. Galls contain one or more developing wasps. Neither the galls nor the insects do much, if any, damage to trees and vegetation. Thus, galls are considered to be only a minor pest problem.

## Control

Control measures against gall wasps are not recommended unless they swarm in great numbers and produce large masses of galls that visibly deface trees and shrubbery. In rare cases where control is indicated, however, apply a residual spray to all affected trees, shrubs, and vegetation, using a high-powered pressure sprayer. The following insecticides are recommended: 1.5% diazinon, Dursban, or Baygon; or 2% malathion, Sevin, Gardona, or thanite.

# OAK MOTH CATERPILLARS

See: *Caterpillars* in this section.

# OAT BUGS

See: *Thrips* in this section.

# ODD BEETLES

*Phylum: Arthropoda    Class: Insecta    Order: Coleoptera*

These tiny beetles, only a fraction of an inch long, are closely related to the carpet beetle. The male and female look entirely different. The male is elongated with the usual splitback beetle appearance, while the female looks like a larva or grub, but has legs and antennae.

These beetles become household pests when they infest garments, muslin, bedding, china closets, book cases, white tissue paper, silk, and woolens.

## Control

Follow the control methods recommended for *Buffalo Bugs* and *Casemaking Moths* in this section.

# ODOR
### (Dead rat, bat, squirrel, mouse, etc.)

The products listed below are recommended to mask, or neutralize, the odor produced by dead rats, mice, bats, birds, squirrels, and other animal pests that occasionally die within walls, subfloors, and other inaccessible areas inside the house. It may be necessary to drill one or more small holes through which to inject deodorant chemical under pressure.

The following deodorants/disinfectants are recommended: neutroleum alpha, Bactine, dutrol, quarternary ammonium compounds, Styronol 1622, zephiran chloride, metazene, Nilodor, and Polycide.

# OLD HOUSE BORERS

*Phylum: Arthropoda    Class: Insecta    Order: Coleoptera*

The old house borer is a common and destructive wood-boring beetle that is a serious pest in houses and buildings because it attacks and damages the structural wood and timbers. This beetle is black to dark brown with gray pubescence and a white patch on its back. The old house borer is a serious European pest in Germany, Norway, and Denmark. In the United States, it lives in the Atlantic Coast states, and also in Mississippi, Louisiana, Texas, New York, and Pennsylvania.

The presence of these beetles may be detected by the rasping or ticking sound made by the larvae as they bore within the wood. Also, the appearance of very fine, powdery sawdust in sapwood or within the tunnels and galleries, when a sharp probing instrument is used, indicates the presence of old house borers. Further evidence is the presence of roughly one-fourth-inch diameter oval holes made on the surface of wood and timber by emerging adult beetles.

This beetle especially likes to attack roof timbers, structural wood, and even wood flooring. The types of wood most likely to be attacked are pine, spruce, hemlock, and fir. Infestation and damage by these beetles may occur over a period of several years.

### Control

Extensive and/or long-standing infestations of old house borers will require the expertise of a licensed pest control operator who is equipped to use fumigation and other difficult control measures. Smaller, more limited infestations, however, can be treated by the average person.

Use a high-pressure liquid sprayer, or duster, to force insecticide through the holes made by the beetles. Force in enough chemical to fill the entire tunnel network within the wood. NOTE: In some cases, it may be necessary to drill additional holes through which to treat the wood. As always, plug up the holes with wooden dowels, caulk, or putty when you have finished treating the wood.

The following chemicals are recommended: 5% PCP (pentachlorophenol), 0.5% lindane, methyl bromide, or a soap-solution of nicotine sulfate.

As a preventive measure against old house borers, spray or brush the surface of wood and timbers with 5% PCP in oil, or Woodtreat-TC.

## ONION THRIPS

*Phylum: Arthropoda    Class: Insecta    Order: Thysanoptera*

The onion thrip is a tiny but common and widely distributed pest that attacks onions, tobacco, beans, and various other plants. This thrip is pale yellow to brownish in color, and is about one twenty-fourth of an inch long.

**Control**

Treat affected plants with a residual spray or dust of: 1.5% Baygon, diazinon, or Dursban; 2% malathion, Sevin, DDVP, or SBP-1382®; or 1% lindane.


# OPOSSUMS

*Phylum: Chordata     Class: Mammalia     Order: Marsupialia*

Opossums are roughly cat-size or larger, with rather tufted and grizzled fur that is grayish to black in color. These animals have a slender, tapering nose and a long, rat-like tail. Opossums occasionally become pests around houses and farms where they forage for food, especially at night. These animals will eat practically anything, including carrion, insects, fish, berries, fruits, vegetables, and meats.


**Control**

Screen or cover all areas under houses, sheds, and buildings, and especially all openings to poultry houses. Foraging opossums may be trapped easily with cage traps, steel traps, or live traps. Recommended baits are fish, dog food, meat scraps, or mixed meat and vegetables.


# ORIENTAL FRUIT MOTHS

*Phylum: Arthropoda     Class: Insecta     Order: Lepidoptera*

The oriental fruit moth is a small, brownish to gray moth that lives throughout the United States. A serious pest of peaches and other fruits, it produces several generations per year. First-generation larvae (caterpillars) bore into young, green twigs on the fruit trees, while later-generation larvae bore into the actual fruit, much as the codling moth does.


**Control**

Use the control methods recommended for *Codling Moths* in this section.

## PANTRY PESTS

Various types of pests occur in pantries and cupboards where they feed on a variety of dried, stored foods. The most common pantry pests are Mediterranean flour moths, Indian meal moths, mealworms, grain moths, cigarette beetles, drugstore beetles, flour beetles, cereal and grain beetles, hide beetles, larder beetles, grain-boring beetles, rice weevils, grain weevils, sawtooth grain beetles, warehouse beetles, cadelles, flat grain beetles, red flour beetles, book lice, bean and pea weevils, grain mites, ants, and cockroaches.

Most of these pests are covered *individually* in this section. However, if you are unable to determine which type of pest is present in your pantry, it is suggested that you follow the *general* pantry treatment method given below.

### Control

Remove and discard all infested flour, grain, cereals, spices, cornmeal, etc. Remove all dishes, cookware, food containers, and shelfpaper from all cabinets, shelves, drawers, cupboards, and pantries. Apply a residual spray, or dust, to all areas listed above, paying particular attention to all cracks, crevices, and other insect hiding places.

Recommended insecticides are: pyrethrins, resmethrin, and DDVP contact sprays; 1.1% Baygon; 0.5% diazinon or Dursban; 0.25% Ficam; 2% malathion or Sevin; 3% methoxychlor; or 1% lindane residual sprays or dusts.

## PEA WEEVILS

See: *Bean Weevils* in this section.

## PELIDONTAS, GRAPE

*Phylum: Arthropoda      Class: Insecta      Order: Coleoptera*

The grape pelidonta is a common and destructive beetle pest. The adult beetle is an inch or more long and looks somewhat like a June beetle, but is yellowish, with three black spots on each wing. The adult beetles do most of their damage to plants and vegetation, while the larvae feed mainly on rotting wood.

**Control**

Follow the control measures recommended for *Japanese Beetles* in this section.

## PENNSYLVANIA WOODS ROACHES

*Phylum: Arthropoda      Class: Insecta      Order: Orthoptera*

The Pennsylvania woods roach is trim and chestnut-brown in color, with the thorax and wing pads bordered in white. The male varies from slightly less than, to over, one inch in length. The female, however, is smaller. These roaches commonly exist throughout the southern, eastern, and midwestern states northward into Canada.

The males have well-developed wings, and can be seen flying about outdoors. Houses located near wooded areas commonly are invaded by these roaches.

**Control**

Treat inside areas where roaches occur with a residual spray, or dust, of: 1% diazinon or Dursban; 1.5% Baygon; 2% DDVP; 5% malathion; or 0.5% Ficam.

## PICKLE WORMS

See: *Caterpillars* (Control) in this section.

## PICNIC ANTS

See: *Ants* (*Outdoor-Nesting*) in this section.

## PIGEONS

*Phylum: Chordata      Class: Aves      Order: Columbiformes*

Pigeons are very common and pervasive pests of mankind throughout the United States and many parts of the world. These birds mate for life, which may reach fifteen years or more. They have a high reproductive capacity, and they breed year round. Pigeons feed mainly on grain and seeds, but also will eat some green food and fruit.

Pigeon movement usually is extensive, and involves feeding areas, rest sites, and nesting areas. Such diversity of movement makes them quite difficult to control.

Pigeon roosts that have been in use for three years or more should be avoided because of the danger of *histoplasmosis*, a respiratory disease that affects humans. Spores that cause this disease are produced in the soil beneath roosts which is enriched by pigeon droppings, and are blown into the surrounding air where they may be breathed in by humans.

### Control

Pigeons are difficult to control, and attempts by individuals may not prove very effective. Begin, however, by removing all grain, seeds, food, and water from all feeding areas used by pigeons. Bird-proof all buildings frequented by pigeons, using screen wire or other suitable material. Set cage traps, and eliminate all captured pigeons. If shooting is allowed, use this method to reduce the number of pigeons. Finally, poison baits can be used, but this must be done legally, with great care, and with close supervision because of the danger to protected bird species, animals, pets, and children. NOTE: Pre-baiting is necessary for a few days before poison bait is applied.

Some recommended poisons are: 1% Strychnine on grain; DRC-1339® (starlicide); and Avitrol. Also recommended are: Rid-A-Bird perches, Roost No More, For The Birds, endrin, fenthion, Ornitrol, and Mesurol.

### PILLBUGS

*Phylum: Arthropoda    Class: Crustacea    Order: Isopoda*

Pillbugs are crustaceans rather than insects. They live both in Europe and the United States, and are found in and around houses where there is excessive moisture combined with decaying organic matter and vegetation. Without excessive moisture, pillbugs soon perish.

Pillbugs may injure young plants and/or the roots of these plants, especially where there is a heavy infestation. Occasionally, moreover, they invade houses, but are harmless and do no damage. They soon perish after leaving their protective high-moisture habitats.

### Control

Remove boards, planks, lumber, paper, cardboard, bricks, decaying vegetation, and other trash and debris from yards, flower beds, mulch areas, etc. Allow sunshine and air circulation to reach damp, moist areas.

Occasionally, however, when pillbugs occur in large numbers and/or damage flowers and plants, it may be necessary to apply chemical control measures. In this case, apply a residual spray around and near buildings, building foundations, damp and moist areas, and crawl spaces beneath buildings and houses.

Recommended insecticides are: 2% malathion or Sevin; 1% diazinon or Dursban; 1.5% Baygon; or follow label directions and apply any of the following chemicals: Mesurol, Zectran, metaldehyde, chlormephos, or bendiocarb.

## PINE MOTHS

*Phylum: Arthropoda      Class: Insecta      Order: Lepidoptera*

These moths attack and injure Virginia, pitch, red, yellow, jack, and other pines along the Atlantic seaboard and as far west as New Mexico. Pine moths are generally small, copper-colored moths with approximately a one-half inch wingspan. The eggs, which are yellow, are laid on needles and terminal buds in June. The resulting larvae (caterpillars) are yellow and about one-half inch long. They mine into buds, spin webs around needles, and finally burrow down into the new-growth twigs. Second generation adults appear in August. Larvae are produced again and spend the winter as pupae, often in the soil.

### Control

In May, apply to the bark and main trunk of pines any of the following insecticides: endosulfan, trichlorofon, or dimethoate. Follow label directions for correct strength in each case.

## PLANT BUGS

*Phylum: Arthropoda      Class: Insecta      Order: Hemiptera*

This is a large family of bugs that live on vegetation throughout much of the world. These are true bugs, with long needle-like beaks through which they suck plant juices, often doing serious damage. One example is the tarnished plant bug that is about one-fourth of an inch long and varies from pale brown to black in color. Another example is the four-lined plant bug which is yellowish-green, with four longitudinal black lines on the back.

**Control**

When these bugs are visible, or when leaf curling or wilting occurs, apply a contact or residual spray of: Sevin, trichlorofon, malathion, sabadilla, diazinon, or Gardona. Follow label directions for exact strength and application instructions.

## PLANT HOPPERS

*Phylum: Arthropoda      Class: Insecta      Order: Homoptera*

Plant hoppers are a large group of plant pests, but are somewhat less abundant than are leafhoppers and froghoppers. Most plant hopper species in the United States vary from slightly less than, to about, one-half inch in length. These insects feed on the foliage of trees, shrubs, herbaceous plants, and grasses. Few of them, however, cause serious damage to cultivated plants.

**Control**

Whenever control measures against plant hoppers seem necessary, follow the recommendations given for *Leafhoppers* in this section.

## PLANT LICE, JUMPING

See: *Potato Psyllids* in this section.

## POCKET GOPHERS

See: *Gophers* in this section.

## POMACE FLIES

*Phylum: Arthropoda      Class: Insecta      Order: Diptera*

Pomace flies, also known as fruit flies and/or vinegar flies, are very small pests that can pass through ordinary fly screens on doors and windows. These tiny flies are common in homes, restaurants, fruit markets, warehouses, canneries, and similar areas. They are attracted to both human and animal feces, fruits, vegetables, and other uncooked foods. Grapes and other fruits that have been contaminated by pomace flies may cause diarrhea or other gastrointestinal disorders when consumed by humans.

### Control

Locate and destroy all rotting or fermenting fruits and vegetables. Keep all open fruit and vegetable containers in the refrigerator. Eliminate all empty fruit cans, vegetable cans, ketchup bottles, etc.

If flies still persist inside the house, look for scraps of fruit, peels, or other food bits under the stove, cabinets, and appliances. If there are cracks or crevices in linoleum or wood floors, treat these with a residual spray, including baseboards and moldings.

The following aerosol or space sprays are recommended for flying adults inside the house: 1% DDVP (or DDVP resin strips); or 1–2% SBP-1382®, resmethrin, allethrin, or pyrethrins.

For floors, cracks, crevices, walls, ceilings, etc., the following residual sprays are recommended: 1% dimethoate; 3% malathion; or 1% ronnel or Rabon.

## POULTRY BUGS

*Phylum: Arthropoda*      *Class: Insecta*      *Order: Hemiptera*

The poultry bug is a member of the true bug group known as the *Hemiptera*. Like others of this group (including the bedbug, assassin, and conenose bugs) the poultry bug is equipped with a long needle-like beak through which it sucks blood. Although it resembles the bedbug, the poultry bug has longer legs and a longer beak, and is more active. Unlike the bedbug, the poultry bug does not give off an odor.

The poultry bug exists throughout the southwestern United States and Mexico, where it is a severe pest of poultry, sucking their blood, especially at night. This bug also will invade houses where it may bite the human inhabitants.

### Control

Treat poultry houses and cages thoroughly—especially the floors, walls, and roosts, with a residual spray. For poultry bugs inside the house, use a DDVP or pyrethrin contact spray for the quick knock-down of all visible bugs. For bugs present indoors, but not visible, use a residual spray to treat floors, walls, cracks, crevices, baseboards, moldings, etc.

The following residual sprays are recommended: 0.5% Dursban or diazinon; 3% Sevin or malathion; 1.5% Baygon; 1–2% DDVP, pyrethrins, or trichlorofon; or 1% ronnel or lindane.

## POTATO PESTS

Generally, potato pests fall into the following groups: aphids, potato beetles, leafhoppers, psyllids, scab gnats, and caterpillars (cutworms, cankerworms, etc.). Most of these insects are destructive to potato plants, and should be eliminated as soon as they are detected.

### Control

Try to determine which type of pest is present on your potato plants, then turn to the appropriate description in this section. For example, if potato aphids are present, turn to *Aphids* for control measures. If potato beetles are present, turn to the *Colorado Potato Beetles* for appropriate control recommendations. If worms or caterpillars are present, turn to *Caterpillars, Cutworms, Cankerworms,* etc. for recommended control measures.

## POTATO PSYLLIDS

*Phylum: Arthropoda      Class: Insecta      Order: Homoptera*

These very small insects, only a fraction of an inch long, somewhat resemble miniature cicadas. Also called "jumping plant lice," potato psyllids are similar to aphids, but have strong jumping legs. The potato and/or tomato psyllid transmits a virus that causes "yellows" in potatoes, tomatoes, peppers, and eggplants. These insects also suck juices from plants, thus multiplying their damage.

### Control

Follow label directions and apply any of the following insecticides: Temik, Furadan, Systox, Di-Syston, Monitor, Azodrin, Orthene, Thimet, pirimicarb, or rotenone.

## POWDER-POST BEETLES

*Phylum: Arthropoda      Class: Insecta      Order: Coleoptera*

Powder-post beetles are very destructive pests of wood, and are second only to termites in this respect. Adult powder-post beetles are very small, varying in size from a fraction to about one-fifth of an inch in length. They appear flattened, and are reddish-brown to black in color. The adult female deposits her eggs within the pores of wood where they hatch into larvae. The larvae, in turn, bore throughout the wood until the adult stage is reached. The adult beetles then bore holes in the surface of the wood and emerge.

A powder-post beetle infestation can be detected by the presence of extremely fine powder, almost like flour, which falls from the surface holes in the wood. Other woodboring insects leave a coarser, sawdust-like powder, along with fecal pellets in most cases. Powder-post beetles typically infest the sapwood of hardwood, and may attack hardwood flooring, tool handles, furniture, and the structural timbers within houses and buildings. Ash, oak, elm, pecan, and walnut are favorite woods of these beetles.

### Control

Through the holes made by the beetles in the wood surface, apply a residual spray or dust under pressure sufficient to reach and thoroughly treat the entire tunnel network within the wood. In some cases it may be necessary to drill additional holes in order to completely treat all infested timbers and wood.

When treating flooring or furniture, it is best to use a light oil solution (such as mineral oil) of insecticide. First, however, apply a small amount to see whether staining or discoloration occurs. If so, switch to another oil carrier, or perhaps to water.

For large surface areas, use a fan nozzle on your sprayer, or brush on the insecticide with a paint brush. NOTE: Long-standing infestations of these beetles that have spread throughout the entire structural framework of the house, and also within the walls, will likely require large-scale fumigation by a licensed pest control operator.

Recommended pesticides for powder-post beetles are: 5% PCP (pentachlorophenol); 0.5% lindane; 1.5% Baygon; or 1% diazinon or Dursban.

Recommended repellents are: Woodtreat-TC and 5% PCP. Spray or brush these chemicals onto all wood surfaces that you wish to protect from woodboring beetles.

## POWDER-POST TERMITES

*Phylum: Arthropoda    Class: Insecta    Order: Isoptera*

The powder-post termite lives only in the southern coastal areas from Florida westward to Louisiana. Found only within buildings, this termite is very destructive to the wood it attacks. It completely devours the interior of the structure, leaving only a paper-thin shell of wood or paint at the surface. This pest will attack both woodwork and furniture.

The powder-post termite can be distinguished from drywood termites and powder-post beetles by inspecting closely the infested wood. The powder-post termite produces very tiny holes on the surface, as opposed to the larger holes of beetles. And, unlike the very clean tunnels of the much larger drywood termite, the smaller powder-post termite leaves its tunnels filled with a very fine powder, along with tiny fecal pellets.

### Control

For furniture infestation, use a hypodermic syringe to inject insecticide into the tiny holes on the surface of the wood. Force into the holes enough chemical solution to thoroughly treat the entire tunnel network within the wood.

To treat structural woodwork and timbers, drill one-half inch holes in larger timbers and smaller holes in smaller timbers, then use a pressurized sprayer or duster to force insecticide thoroughly throughout the tunnel network within the timbers and wood.

5% PCP (pentachlorophenol) is recommended for the control of powder-post termites. After the treatment is completed, cover all holes with putty or caulk.

## PUBIC LICE (Crab Lice)

See: *Crab Lice* in this section.

## PUNKIES

See: *Midges* (*Biting*) and *Black Flies* in this section.

## PURPLE MARTINS

See: *Blackbirds* (Control) and *Pigeons* (Control) in this section.

# RABBITS

*Phylum: Chordata    Class: Mammalia    Order: Lagomorpha*

Rabbits frequently do extensive damage to ornamental plants, vegetables, gardens, truck patches, and even trees. Active year round, rabbits feed mostly on leaves, stems, buds, and the bark of trees. They tend to feed in the early morning and late afternoon.

## Control

Rabbits live in heavy vegetation, brush, debris piles, and similar areas. For long-term control of rabbits, clean up these areas if they exist near your garden, plants, trees, etc. Secondly, construct wire fencing around gardens, yards, flower beds, trees, and individual plants. Thirdly, apply naphthalene and/or Arasan as repellents for rabbits. These should be applied liberally on and near plants, vegetables, flowers, trees, etc. that are subject to attack by rabbits. Trapping rabbits with box traps baited with apples, lettuce, or carrots also is recommended. Shooting may be employed wherever it is legal. Finally, live traps may be used.

# RACCOONS

*Phylum: Chordata    Class: Mammalia    Order: Carnivora*

Raccoons live throughout most of the United States. They are very intelligent and persistent animals, and are excellent climbers. Raccoons are active at night, and tend to remain in their dens (located in hollow logs, trees, burrows, etc.) during the day. Raccoons will eat almost anything eaten by man. Often raccoons become pests when they forage for food in outbuildings, attics of houses, garbage cans, chicken houses, and cornfields.

## Control

Trapping is the recommended method to control raccoon pests. Use wire traps, box traps, or live traps baited with fish, chicken, or corn on the cob. NOTE: Wild raccoons often transmit rabies. *Never* touch or handle them with your bare hands.

# RATS
## (Norway, Brown, House, Wharf, or Sewer)

*Phylum: Chordata*     *Class: Mammalia*     *Order: Rodentia*

The Norway rat is the larger of the two rat species that exist in the United States. (The gray-bellied or roof rat is covered separately in this section.) Unlike its smaller relative, the roof rat, the Norway or ground rat lives mainly at ground or floor level. This very common and persistent rat measures seven to ten inches in length, with the tail adding another six to eight inches. The fur is a coarse brownish-gray, with a few scattered black hairs and a gray to yellowish belly. Norway rat droppings may be three-fourths of an inch long, and are capsule-shaped. Numerous droppings over a large area indicate a large rat infestation.

Norway rats live nine to twelve months, become sexually mature at three to five months, and produce up to seven litters of young in a lifetime. These rats may live either near man or some distance away. Frequently they live in ground burrows beneath houses, buildings, lumber piles, trash piles, garbage dumps, stones, or in stream banks. Generally Norway rats eat more meat than do the smaller roof rats. Favorite foods of the Norway rat are corn, flour, wheat, bread, beans, livestock feed, fresh fruits, vegetables, sugar, and even flower bulbs.

These rats may invade houses and apartments when large infestations occur in areas of poor community sanitation.

### Control

Rats will *not* linger where there is neither food nor water.

Effective rat control always involves 4 basic steps: (1) eliminate all rat shelters and harboring areas if possible; (2) remove all food and water supplies; (3) if possible, rat-proof all infested buildings; and (4) kill as many rats as possible.

Practice good sanitation and keep all areas clean of scrap food, vegetables, fruits, grain, animal food, etc. Cover or remove all open or standing water. Keep garbage in cans with tightly-fitting lids. Eliminate all rubbish, trash, and debris. Cut down grass, weeds, and vegetation around houses and buildings.

Norway rats may be killed with poison baits and/or fumigation. In most cases, however, fumigation must be done by a licensed pest control operator. Smaller, limited infestations, on the other hand, may be tackled by the average person. Apply large amounts of a liquid fumigant to all visible rat burrows, then seal off the openings with tightly-packed dirt. If carbon monoxide (automobile exhaust) gas is used, pump it into all rat burrows for twenty to thirty minutes, then seal off the openings with packed dirt.

In recent years, rats have shown some biological resistance to the anti-coagulant poisons. Consequently, these rodenticides may not prove fully effective in every case. Nevertheless, the anti-coagulants are the baits of choice and should be tried first. If they seem ineffective, simply switch to one of the recommended non-anticoagulants listed below.

When rat burrows that are to be fumigated by you are beneath a house or building that is occupied by people, first vacate the building of all human inhabitants before fumigating. This will eliminate the danger of fumigant gas diffusing upward into the building.

Recommended anti-coagulant poisons are: warfarin, Fumarin, Fumasol, Pival, Pivalyn, diphacinone, chlorophacinone, and PMP.

Non-anticoagulant poisons are: ANTU®, arsenic trioxide, zinc phosphide, phosphorus, strychnine, norbormide, and Red Squill.

Recommended fumigants are: carbon tetrachloride, methyl bromide, and carbon monoxide.

NOTE: Norway rats are very suspicious, and do not accept new bait or food sources readily. Pre-baiting, and patience, are necessary when poisoning rats. Follow label directions when applying any of the listed bait-poisons.

## RATTLESNAKES

*Phylum: Chordata     Class: Reptilia     Order: Squamata*
*Suborder: Serpentes*

Rattlesnakes exist only in the Western Hemisphere, with fifteen species native to the United States. Rattlesnakes have a broad, triangular head, a narrow neck, vertical eye pupils ("cat eyes"), and rattles on the tail. All rattlesnake species are poisonous. Unlike many other snakes, rattlers do not lay eggs but, rather, give live birth to their young.

NOTE: A severed rattlesnake head is still dangerous for at least twenty minutes and sometimes longer.

### Control

Wherever snakes exist in large numbers there will be an abundant food and water supply. Eliminate rats, mice, and other rodents from on and near your property. Clean up areas where snakes live, including trash and debris, woodpiles, thick grass or vegetation, compost piles, etc.

Snakes enter houses and buildings near ground level, or below. Check your house foundation all around, and seal off all openings, including very small ones.

If a snake gets inside your house and goes into hiding, place wet cloths in the room where it is thought to be. Cover this pile of wet material with *dry* cloth. Since snakes are attracted to moisture, this method often will lure a snake from hiding, thus allowing you to kill or remove it. Also, an aerosol insecticide bomb released in the room may prove effective in dislodging the snake from hiding.

Snakes outside may be controlled to some degree by poisoning and/or fumigation of their ground burrows. A recommended snake poison consists of: 1 part nicotine sulfate in 240 parts water; pour this solution into a *shallow* pan and cover it with a screen stapled to four pegs, one on each corner. Leave a narrow opening between the pan and screen which snakes may enter, yet which keeps out other animals.

Snakes within ground burrows may be fumigated with: carbon tetrachloride, methyl bromide, or carbon monoxide. If a liquid fumigant is used, pour large amounts of it into all visible snake burrows, then seal off each one with tightly-packed dirt. If carbon monoxide or another gas is used, pump it into the burrows for twenty to thirty minutes, then seal off the holes with tightly-packed dirt.

## RED FLOUR BEETLES

*Phylum: Arthropoda     Class: Insecta     Order: Coleoptera*

The red flour beetle is an important pest of stored foods in grocery stores and in homes. These are very small, reddish-brown beetles that are about one-seventh of an inch long. They live in flour, cereals, spices, and other stored foods in pantries and cupboards.

### Control

Discard all infested flour, cereals, meal, spices, and other foods. Remove all food containers, dishes, cookware, and shelfpaper from all cabinets, cupboards, pantries, shelves, and drawers. Treat these areas with a residual spray or dust of: 1.1% Baygon; 0.5% lindane or Dursban; 1% diazinon; 2% malathion, Sevin, DDVP, or SBP-1382®; or 0.25% Ficam.

# RED-LEGGED HAM BEETLES

*Phylum: Arthropoda    Class: Insecta    Order: Coleoptera*

The red-legged ham beetle is the most important of all the insects that infest dried or smoked meats. The adult beetle is shiny green, or greenish blue, with reddish-brown legs. It varies from about one-seventh to one-fourth of an inch long, and has many tiny "pits" or puncture marks on its back.

Larvae of this beetle bore into meat and do most of the actual damage, while the adult beetles feed on the surface. The red-legged ham beetle is known to infest fish, cheese, bones, hides, herring, dried egg yolks, dried figs, palm-nut kernels, and other foods.

### Control

Install tightly-fitting doors and small-mesh screens on the windows of all rooms where meats, fish, cheese, hides, and similar items are stored. Dispose of badly-infested foods, then rinse all non-infested or lightly-infested foods with clear water. Cover these foods with plastic, cloth, paper bags, or wrapping. Scrub the room thoroughly with hot water and detergent, then apply a residual spray to walls, floors, tables, storage shelves, bins, etc. Wait until the spray has dried completely before returning food material to the room.

For small infestations of red-legged ham beetles that may occur occasionally in kitchens or pantries, follow the control procedures given for *Pantry Pests* in this section.

Recommended residual sprays are: 1.1% Baygon; 0.5% Dursban or lindane; 1% diazinon; 2% malathion or Sevin; or 0.25% Ficam.

# RED SQUIRRELS

See: *Squirrels* in this section.

# RELAPSING FEVER TICKS

See: *Soft Ticks* in this section.

## RICE MOTHS

*Phylum: Arthropoda     Class: Insecta     Order: Lepidoptera*

This pest is of minor importance in most cases, and tends to live mainly in warm, humid climates. The adult moth is pale brownish, with a wing span of about six-tenths to one inch. The larva is dull white, with a dark brown head. It has fine hairs on its body and spins a thick cocoon. The rice moth may do considerable damage to stored rice, chocolate materials, cocoa, nut meats, and some other foods.

### Control

Large quantities of stored rice, or other materials infested with this moth, will require fumigation by a licensed pest control operator for effective control.

Occasional household infestations that occur in pantries, cupboards, and storage rooms, however, can be treated by the average person. Discard all infested food, grain, cereals, cocoa, etc. Remove all dishes, cookware, food containers, and shelf paper from all cabinets, cupboards, pantries, drawers, and shelves. Treat these areas with a residual spray or dust of: 1.1% Baygon; 1% diazinon; 0.5% Dursban or lindane; 2% malathion, Sevin, or resmethrin; or 0.25% Ficam. NOTE: Thoroughly treat all cracks, crevices, and other hiding areas with the residual spray.

## RICE WEEVILS

*Phylum: Arthropoda     Class: Insecta     Order: Coleoptera*

The rice weevil is a pest primarily in warm countries and in the southern United States. The adult rice weevil is reddish-brown, with four faint reddish or yellowish spots on its back, and a long snout. This weevil is tiny, only a fraction of an inch long, and is smaller than the granary weevil. Rice weevils may infest beans, nuts, cereals, wheat, grapes, apples, and pears.

### Control

For both grain-storage infestations and household infestations, follow the control recommendations given for *Rice Moths* in this section.

## ROOF RATS

See: *Gray-Bellied Rats* in this section.

## ROOT MAGGOTS

*Phylum: Arthropoda    Class: Insecta    Order: Diptera*

Root maggots are destructive pests of several different vegetables, fruits, and crop plants. All true root maggots are the larvae of various flies, the adults of which lay their eggs on or in the soil. The resulting larvae (maggots) feed upon the roots of various plants, often causing wilting, stunting, yellowing, and even death.

Some common root maggots are the sugarbeet root maggot, the cabbage root maggot, and the squash root maggot.

### Control

Follow label directions and apply to infested soil any of the following insecticides: Temik, Amaze, Furadan, Di-Syston, malathion, Mocap, ethion, Dasanit, Thimet, or Counter.

## ROOT WEEVILS

*Phylum: Arthropoda    Class: Insecta    Order: Coleoptera*

Several species of root weevils attack various wild and cultivated plants and feed on the foliage. Usually this weevil produces only one generation per year. These pests feed at night and hide in the soil around plants during the day. Although they do no damage, adult root weevils often enter houses in late summer or early fall. They may prove annoying when they crawl over floors and hide in rugs, carpets, and other areas of the house.

### Control

Effective control of root weevils must begin outdoors because these insects always move from infested plants to houses. For both outdoor soil-and-plant treatment and indoor treatment, apply a residual spray of: 1.5% Baygon; 1% diazinon or Dursban; 1% lindane; 2% malathion or Sevin; or 0.25% Ficam.

NOTE: Carefully treat the house or building foundation all around including windows, doors, steps, porches, and crawl spaces. Wall voids, subfloors, and other inaccessible areas should be treated with 5-10% Sevin dust blown in under pressure.

## ROSE CHAFERS

*Phylum: Arthropoda*    *Class: Insecta*    *Order: Coleoptera*

The rose chafer is a slender, tan, long-legged beetle that is closely related to the June beetle, May beetle, or June bug. The rose chafer feeds on the flowers and foliage of roses, grapes, and other plants, including peaches and some other fruits. Rose chafer larvae are small white grubs that live in the soil, often damaging the roots of various plants. Poultry that eat these beetles and grubs tend to become very sick, and often die.

### Control

To control the adult chafers, treat all rose bushes, trees, shrubs, and plants with a residual spray of: 1% lindane; 2–3% malathion; 1.5% Baygon; 1% Dursban or diazinon; 2–3% Sevin; or 0.5% Ficam.

For larva (grub) control, treat the soil around plants, bushes, shrubs, etc. with any of the following soil insecticides: Temik, Amaze, Furadan, Di-Syston, malathion, Mocap, ethion, Dasanit, or Thimet.

## RUSTY GRAIN BEETLES

*Phylum: Arthropoda*    *Class: Insecta*    *Order: Coleoptera*

This small, rusty-reddish beetle is quite resistant to cold temperature, and commonly exists in the northern United States. It attacks stored grain and various stored foods, and may be found under the bark of trees.

### Control

Large quantities of grain or stored food infested by this beetle will require fumigation by a licensed pest control operator for effective control.

For occasional household infestations in pantries, cupboards, cabinets, etc. follow the control measures recommended for *Red Flour Beetles* in this section.

# SADDLEBACK CATERPILLARS

See: *Caterpillars* in this section.

# SAND FLIES

*Phylum: Arthropoda     Class: Insecta     Order: Diptera*

Sand flies are very small to minute, usually hairy and mothlike, flies that occur in moist, shady areas. Generally these flies are found in the southern United States and in the tropics. They are bloodsuckers, and are known to transmit several diseases in various parts of the world. Adult sand flies often swarm around sewers, drains, and garbage areas. The larvae, however, live in decaying plant matter, mud, moss, and brackish water.

### Control

Outdoor control of these flies is difficult because of their range and breeding areas.

For indoor treatment, follow the control measures given for *Drain Flies* and *Filter Flies* in this section.

# SAP-SUCKING INSECTS

There are many types and varieties of sap-sucking insects. These pests feed on the juices and sap of plants, vegetables, flowers, trees, and other vegetation. They often cause stunting, curling, wilting, yellowing, and other conditions in plants. Moreover, some sapsuckers transmit various plant diseases.

### Control

If possible, identify the type of sap-sucking pest before applying control measures. However, if you are unable to determine the type of pest involved, proceed on the principle that most sap-sucking insects can be controlled with soil-applied plant systemic insecticides.

Some recommended plant and soil systemics are: Temik, Amaze, Meta-Systox-R, Di-Syston, demeton, Bidrin, Cygon, and Dimecron. NOTE: Follow label directions when applying any of these chemicals.

## SAWFLIES

*Phylum: Arthropoda*    *Class: Insecta*    *Order: Hymenoptera*

Sawflies are not true flies at all but, rather, belong to the Hymenoptera (wasp) group. Most sawflies are plant-eaters, and tend to feed on the leaves of various plants and vegetation. The female sawfly has a long saw-like ovipositer (egg-depositing device) which is used to insert eggs into the tissues of plants.

The most important sawfly species are discussed individually below.

*Elm Sawfly*—This is a rather large, robust insect that rather resembles a bumblebee. The elm sawfly is dark blue, and measures three-fourths to one inch in length. The female has four small yellow spots on each side of the abdomen. The elm sawfly produces one generation per year, and spends the winter as a larva within the ground. Adults appear in early summer.

*Conifer Sawfly*—This is a medium-sized sawfly that feeds on conifers, and sometimes causes serious damage. Although rare in the midwestern states, conifer sawflies are fairly common in other sections of the United States.

*Common Sawfly*—This is a very large group of common, wasp-like sawflies that are small to medium-sized, and often brightly colored. They live on foliage and flowers. The larvae also feed on plant foliage, and tend to curl part of the body over the leaf when feeding. These larvae feed on various trees, shrubs, and plants, and some are quite destructive. Some common species in this group are:

1. *Larch Sawfly*—This is a very destructive pest of larch. The larvae, when numerous, may cause extensive defoliation of trees over large areas.

2. *Currant Worm*—This "worm" is the larva of a common sawfly, and is a serious pest of currants and gooseberries.

3. *Leafminers*—Many of these are the larvae of various common sawflies. For example, the birch leafminer makes mines in birch leaves, and the elm leafminer attacks elm trees.

### Control

Sawflies and their larvae can be controlled with the same recommendations given for *Caterpillars* in this section.

# SAW-TOOTHED GRAIN BEETLES

*Phylum: Arthropoda     Class: Insecta     Order: Coleoptera*

The saw-toothed grain beetle is a very common pest of cereals, flour, dried meat, macaroni, dried fruits, and chocolate. The adult beetle is very small (about one-tenth of an inch long) and brownish in color. Each side of the thorax has six saw-toothed projections, hence the name "saw-toothed" grain beetle. Because of its small size, this pest easily enters packaged cereals and other foods through tiny cracks and openings. A large infestation of these beetles usually occurs if they are not controlled effectively.

### Control

Discard all infested food, cereals, fruits, etc., including unopened boxes and containers. Remove all dishes, cookware, and shelfpaper from all cabinets, pantries, cupboards, drawers, and shelves. Treat these areas thoroughly with a residual spray or dust, paying close attention to very small cracks, crevices, and other hiding places.

The following insecticides are recommended: 2% malathion or Sevin; 1% lindane; 1.5% Baygon; 1% diazinon; 0.5% Dursban; 0.25% Ficam; or drione dust.

If beetles are visible, use a contact or space spray of DDVP, pyrethrins, or resmethrin to kill them quickly. In any case, however, apply the residual treatment described above.

# SCALE INSECTS

*Phylum: Arthropoda     Class: Insecta     Order: Homoptera*

Scale insects are of two types: armored and soft. The armored scales make up the largest group of these pests, and the scale covering these insects is formed of wax. These scales suck plant sap and juices; and, when numerous, they may kill plants. Armored scales feed chiefly on trees and shrubs, and some are serious pests of orchard and shade trees. Some important armored scales are discussed individually below.

*San José Scale*—This pest appears throughout the United States where it attacks orchard trees, shade trees, and ornamental shrubs. This scale is generally circular in shape.

*Oystershell Scale*—This scale is so named because its shell resembles that of an oyster. This scale exists throughout most of the United States and in many parts of the world and attacks fruit trees, ornamental trees, and shrubs.

*Scruffy Scale*—This scale is very common, whitish in color, and attacks a variety of trees and shrubs.

*Rose Scale*—This insect is reddish with a white scale. A heavy infestation of this scale causes trees and plants to look whitewashed. This pest attacks various berries and roses.

*Pine Needle Scale*—This very common scale attacks pine and other evergreen trees.

The soft scale group also contains a number of injurious pests. The *black scale* attacks citrus and other plants in the southern United States. The *hemispherical scale* infests ferns and other ornamental plants in both homes and greenhouses. And the *cottony maple scale* produces a cotton-like mass at one end.

### Control

Scales are difficult to kill, and several applications of insecticide at three- to four-week intervals may be required for good control of these pests.

For small infestations (when small plants are involved, or only a few plants are infested) dip the plants in a malathion-mineral oil solution or a malathion-detergent-water solution.

For larger infestations, apply an insecticide solution in the fall after growth has stopped, or in the early spring before buds open. The spray also should be applied as the insects hatch.

The following insecticides are recommended for scales: ethion (oil); diazinon (wettable powder or emulsion); Sevin (wettable powder or soap-water emulsion); acephate/Orthene (emulsion); malathion (oil, wettable powder, or detergent-water solution); azinphosmethyl/Guthion (wettable powder); Meta-Systox-R (emulsion); or Dursban (emulsion). NOTE: Because the scale covering is formed of wax, a water-based insecticide will not penetrate this material and, thus, is not effective. You must, therefore, use an oil solution, wettable powder, or emulsion (such as: detergent-water-insecticide).

## SCAVENGER BEETLES

*Phylum: Arthropoda*    *Class: Insecta*    *Order: Coleoptera*

Scavenger beetles are very tiny reddish-brown insects, varying from one-twenty-fifth to about one-tenth of an inch long. These beetles sometimes swarm in large numbers inside homes and buildings, thus becoming quite annoying. They feed on mold and mildew, and tend to live wherever there is persistent dampness, moisture, mildew, and mold. Scavenger beetles are harmless and do no damage.

### Control

Eliminate all dampness, moisture, mildew, and mold. In some cases this action alone is sufficient to control these beetles. However, when a large infestation is present, it may become necessary to use an insecticide. In this case, follow the recommendations given for *Boxelder Bugs* (Control) in this section.

## SCORPIONS

*Phylum: Arthropoda      Class: Arachnida      Order: Scorpionida*

Most readers will be able to recognize scorpions on sight. These arachnids resemble crayfish, but have a long tail with a stinger on the end that curls high over the back. Scorpions are common in all of the southern United States, Mexico, and South America. Most are active at night and hide under bark, rubbish, boards, and stones during the day. Scorpions also enter houses where they become serious pests, hiding in closets, shoes, folded clothing, and similar areas.

Most scorpions are only mildly poisonous, and produce no greater effect than a bee or wasp sting. However, two very poisonous species, capable of causing death to humans, exist in southern Arizona, adjacent areas of California, and in New Mexico and Texas. These scorpions may reach two to three inches in length.

Scorpions feed on small spiders and insects.

### Control

Outdoors, remove all trash, debris, loose boards, stones, bricks, stacked wood, lumber, etc. Treat the entire area where scorpions occur with a residual spray of: 1.5% diazinon; 1% Dursban; or 2–3% Sevin. Recommended dusts are: 2–3% diazinon; 10% Sevin; or 1–2% pyrethrins.

Indoors, treat cracks and crevices in woodwork, baseboards, closets, around water pipes and plumbing, beneath sinks, and around doors and windows. Use any of the residual sprays listed above.

## SCREW-WORM FLIES

*Phylum: Arthropoda      Class: Insecta      Order: Diptera*

The screw-worm fly is a very serious pest in the southern and southwestern United States. These flies are true parasites that lay their eggs in the open sores of animals and man. The larvae (maggots) hatch and attack the living tissue surrounding them. These flies also occur as pests inside houses in the southwest.

The screw-worm fly is a member of the large group known as blowflies. The adult screw-worm fly is a deep greenish-to-blue metallic in color, with a yellow, orange, or reddish face. It is further characterized by three dark stripes on the thorax.

### Control

Outdoor control of these flies, and the protection of animals attacked by them, is very difficult. Thus, only indoor control measures are given here.

For indoor control, see *Flying Insects* (Control) in this section.

## SEED TICKS

*Phylum: Arthropoda*    *Class: Arachnida*    *Order: Acarina*

"Seed" ticks are simply the immature (nymphal) forms that develop into adult ticks.

### Control

See: *American Dog Ticks* and *Brown Dog Ticks* in this section.

## SEWAGE FLIES

See: *Drain Flies* in this section.

## SHEEP KEDS (Sheep Ticks)

*Phylum: Arthropoda*    *Class: Insecta*    *Order: Diptera*

The sheep ked or sheep tick is a member of the louse fly group, and is a worldwide parasite of sheep and goats. This pest is a wingless, bloodsucking species that varies from slightly less than, to slightly more than, one-fourth of an inch in length. The head of the sheep ked is short and appears sunken into the thorax. The body of this pest is leathery, spiny, and sack-like in appearance.

### Control

Follow label directions and apply to affected animals a spray or dip of: coumaphos, dioxathion, lindane, malathion, pyrethrins, ronnel, rotenone, toxaphene, or diazinon.

## SILVERFISH

*Phylum: Arthropoda*    *Class: Insecta*    *Order: Thysanura*

Silverfish and firebrats are very similar in appearance. These very common household pests are long and slender, but flattened. They are broad at the front, but taper gradually to the rear rather like fish. The silverfish adult is about one-half inch long, silvery in color, and with long, slender antennae. All silverfish are wingless, and develop without metamorphosis. Thus, the young look like the adults, except smaller.

Silverfish may occur anywhere inside the house. However, they are especially fond of starchy material, paper, cotton, flour, dried beef, rolled oats, synthetic fibers of various types, sugar, glue, and paste.

### Control

Apply a residual spray to all hiding areas inside the house. Pay close attention to all cracks, crevices, baseboards, cupboards, pantries, wall cracks and voids, drawers, wooden moldings, and water and steam pipes. Use a dust to treat wall voids, crawl spaces and other hard-to-reach areas. If a large area, such as an entire attic, is to be treated, a space spray can be used.

Recommended residual and space sprays are: 1–2% DDVP; 1% diazinon; 0.5% Dursban; 0.25% Ficam; 1% resmethrin; or 2% malathion.

Recommended dusts are: 2% diazinon or silica gel; or 3% malathion.

## SKUNKS

*Phylum: Chordata*    *Class: Mammalia*    *Order: Carnivora*

Two species of skunks exist in the United States, the striped skunk and the spotted skunk, with the former being the most common. Skunks are mainly active at night, feeding on grubs, insects, small rodents, garbage, birds and their eggs, fruits, and berries.

Skunks often become pests of mankind because of their nauseating odor, and because they destroy and deface lawns and turfs with their digging for grubs and insects. Skunks also become pests when they kill poultry, and when they den beneath buildings and houses. Moreover, skunks are known to transmit rabies to humans.

### Control

When skunks den beneath a house or building, close off all but one opening. Sprinkle flour in front of this opening and check it after dark. If tracks are present, indicating that the skunks have departed, close off this opening and seal it tightly to prevent further entrance by the skunks. Sheet metal, wire netting, concrete, or lumber is recommended for sealing off these entrances.

For the control of skunks that are defacing lawns or turfs, apply the treatment recommended for *Grubs* in this section.

Skunk odor on your person or clothing should be treated with neutroleum alpha or a solution of vinegar or chlorine bleach in water.

### SLUGS

*Phylum: Mollusca     Class: Gastropoda     Orders:* Various

Slugs are soft-bodied, slowly moving, slimy creatures without a shell. They are closely related to clams and oysters, and often become pests in and around homes, flowerbeds, mulch beds, basements, dooryards, etc. Most slugs are gray or mottled, and they require a high-moisture environment in order to survive. These pests may appear in large numbers in damp basements, crawl spaces, flower and mulch beds, and on patios. Slugs leave a glistening trail behind them wherever they go.

### Control

Slugs cannot survive without dampness and moisture. Remove all damp, moist, rotting wood, boards, trash, and debris. Treat areas where slugs occur, including crawl spaces, with a bait or spray of: metaldehyde (5% bait or 0.5% spray); Zectran (2% bait or spray); or Mesurol (2% bait).

## SMOKEY BROWN COCKROACHES

*Phylum: Arthropoda    Class: Insecta    Order: Orthroptera*

The smokey brown cockroach is similar in appearance to the American cockroach, but is smaller. Slightly over one inch long, it is shining brownish black to mahogany in color. The smokey brown cockroach is common from Texas eastward to the Atlantic Coast, and has been reported in North Carolina, Indiana, and Iowa.

In the South, this cockroach may become a serious household pest, while in the North it tends to occur mostly in greenhouses. This cockroach can enter homes in firewood and other items that have been stored in garages or other outbuildings. These insects also can enter homes through small cracks in the walls or floors. Outdoors, smokey brown cockroaches live under dead leaves, around shrubs, flowers, and trees, and beneath structures built off the ground.

### Control

When a heavy infestation of smokey brown cockroaches occurs, you should treat both indoor and outdoor premises.

Outdoors, treat flowerbeds, leaves, debris, mulch, woodpiles, garages, outbuildings, beneath structures sitting off the ground, around shrubs and trees, and around the foundation, doors, and windows of your home.

Indoors, follow the control procedures recommended for *Cockroaches* in this section. For both indoor and outdoor treatment, use the insecticides recommended for *Cockroaches* in this section.

## SNAILS

*Phylum: Mollucsa    Class: Gastropoda    Orders:* Various

Snails are slowly moving pests with shells. They are related, and similar to, slugs. Snails also require a high-moisture environment for survival. Typically, these pests are grayish yellow to brown in color, and their shells resemble a spiral. Snails are bothersome pests in many parts of the world, particularly in dooryards, gardens, on patios, etc. Like slugs, snails also may occur in damp basements, crawl spaces, flowerbeds, mulch beds, gardens, and on patios and porches. Snails may damage plants when they occur in large numbers.

### Control

Follow the control procedures given for *Slugs* in this section.

## SNAKES

*Phylum: Chordata     Class: Reptilia     Order: Squamata*
*(Suborder: Serpentes)*

Most snakes are non-poisonous and actually are beneficial because of the rodents and insects they eat. These beneficial snakes should not be killed unless they exist in large numbers or otherwise become serious pests. Of the 116 species of snakes found in the United States, only nineteen are poisonous. Fifteen of these poisonous snakes are rattlesnakes, two are moccasins (copperhead and cottonmouth), and two are coral snakes. All poisonous snakes, except coral snakes, have vertical eye pupils and a pit on each side of the head between the nostril and eye. These same snakes also tend to have broad, triangular heads and narrow necks.

Snakes are warm-blooded animals, and are active when the temperature is warm to hot. Many types of snakes become pests by entering houses and buildings, crawl spaces, the undercarriages of automobiles, yards, gardens, poultry houses, trees, bird nests, etc.

### Control

Eliminate food sources such as rodent and insect infestations. Clean up debris, trash, compost piles, woodpiles, and thick grass, weeds, and vegetation.

When a snake enters a home or building and disappears before it can be removed, place a pile of *wet* cloths in the same room and cover this wet material with *dry* cloths. Snakes are attracted to moisture, and usually this procedure will lure the snake from hiding. When it appears, you can kill it or remove it from the house. Releasing an aerosol insecticide bomb inside the room also may be effective in dislodging the snake from hiding.

When snakes exist in large numbers, especially venomous ones, it is advisable to poison them. Make a solution of 1 part nicotine sulfate in 240 parts water and pour this into a number of *shallow* pans. Cover the pans with a screen stapled to four pegs, one on each corner. Leave a narrow entrance between the pan and the screen that will admit snakes but keep out other animals. Place these pans in areas where snakes are most numerous.

Snakes within ground burrows may be fumigated. Use large quantities of carbon tetrachloride or methyl bromide, then seal off the entrances with tightly-packed dirt. Carbon monoxide (automobile exhaust gas) also may be used to fumigate snake burrows. Pump this gas into the burrows for twenty to thirty minutes, then seal off the entrance.

## SOFT TICKS

*Phylum: Arthropoda     Class: Insecta     Order: Acarina*

Ticks are divided into two basic types: hard ticks and soft ticks. Those discussed previously in this handbook are hard ticks, including the very common American dog tick and brown dog tick.

Hard ticks look different than soft ticks. When viewed from above, the mouthparts are visible on the front of hard ticks, but are hidden beneath the anterior margin of the body on soft ticks. Some species of soft ticks attack humans, but are more common on birds, bats, and small mammals. Some important soft tick species are discussed below:

*Fowl Tick*—Often called the "blue bug," this tick lives in the southern United States where it attacks fowl, in particular, and transmits the disease spirochetosis. Occasionally this tick will attack poultry workers.

*Relapsing Fever Tick*—This tick tends to exist in nests and seldom is it seen by humans. It attacks humans, especially in mountain cabins, camping areas, caves, etc., in areas of the country where this tick lives, often transmitting a disease known as relapsing fever. This tick lives in Oregon, Washington, California, Nevada, Colorado, and the southern United States.

*Pajaroello Tick*—This soft tick lives in California where it attacks humans, deer, and other mammals. The bite of this tick is greatly feared by those familiar with it because it, too, transmits relapsing fever.

*Spinose Ear Tick*—This soft tick occurs in the southwestern and western United States where the immature forms (larvae and nymphs) parasitize the ears of horses, mules, cattle, sheep, dogs, cats, and wild animals. This tick rarely attacks humans.

**Control**

Follow the control procedures, including the same chemicals, recommended for *American Dog Ticks* and *Brown Dog Ticks* in this section.

## SOD WEBWORMS

*Phylum: Arthropoda*      *Class: Insecta*      *Order: Lepidoptera*

Sod webworms are the larvae of grass moths. Webworms tend to feed around the base of grass plants where they bore into the stems, crown, and roots. These insects may become serious pests in lawns and turfs.

### Control

Follow label directions and treat affected sods, grasses, lawns, turfs, etc. with any of the following insecticides: ethoprop (granules); Baygon (70% wettable powder); Sevin; Dursban (41% emulsifiable concentrate); diazinon (50% wettable powder, 5% granules, or 48% emulsifiable concentrate); trichlorfon (40.5% emulsifiable concentrate); Proxol or Dylox; or Aspon (13% emulsifiable concentrate).

## SOUTHERN PINE BARK BEETLES

*Phylum: Arthropoda*      *Class: Insecta*      *Order: Coleoptera*

This small beetle, which measures about one-sixth of an inch or less in length, is a very serious pest of pines in the southern United States. These small beetles appear in large numbers and attack all species and ages of pine, often killing trees and causing huge losses.

### Control

Follow label directions and apply 0.5% lindane, 1% Baygon, or 1% Dursban to the bark of affected trees.

## SOUTHERN CORN ROOTWORMS

See: *Corn Rootworms* in this section.

## SOWBUGS

*Phylum: Arthropoda    Class: Crustacea    Order: Isopoda*

Sowbugs are very similar to pillbugs, and both species are world-wide pests. Like pillbugs, sowbugs live in damp, moist soil where they feed on decaying organic matter. They often attack and kill, or damage, seeds and young plants.

Sowbugs occasionally invade homes, but are harmless and do no damage.

### Control

Remove boards, paper, cardboard, decaying vegetable and organic matter, rocks, trash, and debris.

Apply a residual spray to flowerbeds, mulch beds, the lower walls and foundations of houses and buildings, and to all crawl spaces beneath.

The following insecticides are recommended: Sevin; diazinon; malathion; Baygon; Dursban; Zectran; Mesurol; metaldehyde; bendiocarb; and chlormephos.

## SPARROWS

See: *Blackbirds* (Control) and *Pigeons* (Control) in this section.

## SPIDERS

*Phylum: Arthropoda    Class: Arachnida    Order: Araneida*

Spiders are arachnids, not insects. All spiders have eight legs, no antennae, six to eight eyes, and a two-part body. All spiders are carnivorous predators that feed on insects and various other arthropods. Thus, most spiders are beneficial.

Spiders are hardy, and can survive for very long periods without food or water. Most are nocturnal and shy, and those that frequent houses usually prefer cool, dark, undisturbed places in which to nest and hide. Almost all spiders possess poison glands, but most spiders cannot penetrate human skin and, thus, cannot bite. Moreover, the venom of these spiders is not considered dangerous. Consequently, there are only two types of spiders that should be considered dangerous in the United States, the black widow and the brown recluse, both of which are discussed separately in this section.

Spiders may invade homes and build up large colonies. These infestations occur most frequently in closets, attics, basements, and other darkish and secluded areas of the house.

### Control

Clean out spider webs and kill all young and adult spiders that are present. Apply a residual spray or dust to closets, attics, basements, and other darkish, secluded areas where spiders are likely to occur. Treat all cracks, crevices, holes, and other hiding places.

Outside, remove all trash, boards, lumber, debris, etc. Treat outbuildings, the ground, and other areas where spiders occur with the same residual spray or dust used indoors.

The following chemicals are recommended for both indoor and outdoor spider control: 1% diazinon; 0.5% Dursban; 0.25% Ficam; 1.5% Baygon; 1-2% pyrethrins; 0.5% resmethrin; 3% malathion; or 1% ronnel.

## SPIDER BEETLES

*Phylum: Arthropoda     Class: Insecta     Order: Coleoptera*

Spider beetles are very small insects, varying from a fraction to about one-fifth of an inch in length, with long legs and a rather spider-like appearance. These beetles exist world-wide, and are active even in freezing temperatures. They infest cereals, cereal products, grains, dried fruits, meats, wool, hair, feathers, rat and mouse feces, and plant and museum specimens.

### Control

For spider beetles inside the home, in such areas as pantries and cupboards, stored specimens, woolens, etc., follow the control procedures given for *Pantry Pests* and *Hide Beetles* in this section.

For spider beetle infestations in large areas such as warehouses, storage houses, etc., fumigation by a licensed pest control operator is recommended. Smaller, limited infestations, however, may be treated by the average person. In this case, use any of the following residual sprays or dusts: 1.5% Baygon; 1% diazinon and Dursban; 2% malathion; or 1% lindane.

## SPIDER MITES (Clover Mites)

*Phylum: Arthropoda       Class: Arachnida       Order: Acarina*

Like most other mites, spider (clover) mites are very tiny creatures. They are brownish to reddish in color, and are common throughout the United States. These pests attack various plants, grasses, and trees, especially strawberry, sweet pea, sycamore, tomato, violet, apple, apricot, arbor vitae, cherry, clover, dandelion, iris, ivy, peach, pecan, plum, poplar, and others.

Spider mites often become very annoying pests when they swarm by the hundreds, or thousands, beneath windows and doors and cover the floors and walls of homes. Although they do not bite humans, crushing these mites leaves reddish-brown stains on walls, floors, furniture, etc.

### Control

For spider mite infestations of trees and plants, apply any of the following miticides: dicofol (Kelthane) - 35% wettable powder; tetradifon (Tedion) - 25% wettable powder; Morestan - 25% emulsifiable concentrate; Cygon - 23% wettable powder; or Plictran - 50% wettable powder.

For house infestations, apply a residual spray of: 2% malathion, 1% diazinon, 1% Baygon, or 0.5% Dursban to outside walls, windows, porches, and doors. If mites are numerous inside, treat interior walls and inside window sills with any of these insecticides.

## SPITTLEBUGS (Froghoppers)

*Phylum: Arthropoda       Class: Insecta       Order: Homoptera*

These are small insects, rarely over one half of an inch in length, that are very similar to leafhoppers. Spittlebugs are plant feeders, and are so-named because the nymphs produce a frothy, spittle-like mass or substance on leaves, shoots, and stems. Spittlebugs are mostly brown or gray in color, and feed on shrubs and herbaceous plants. Most spittlebugs prefer grasses and plants, but a few attack trees.

### Control

Follow the control procedures given for *Leafhoppers* in this section.

## SPRINGTAILS

Phylum: Arthropoda     Class: Insecta     Order: Collembola

Springtails are very small, primitive insects, only a fraction of an inch long, that develop without metamorphosis. Thus, the young closely resemble the adults. Springtails occur in moist surroundings where they feed on algae, fungi, and decaying vegetation.

Springtails may invade swimming pools and houses in great numbers, thus becoming very annoying pests. They can enter homes through screening on windows and doors, around vent pipes, or through tiny cracks under doors. Inside the home, springtails tend to live in kitchen and bathroom areas where there is water and moisture. They hide in tiny cracks and crevices, within walls, and even within the moist potting soil of house plants.

Springtails do not attack humans, but may crawl over the skin, causing itching and irritation.

### Control

Use a fan or other source of ventilation to air-out and dry areas where there is high moisture and dampness. Use a residual spray or dust to treat walls, baseboards, floors, cracks, crevices, beneath sinks and cabinets, and around bathtubs and showers. Treat wall voids with a dust.

The following insecticides are recommended: 3% malathion; 0.5% diazinon or Dursban; or 1% Baygon. Use the dust form of these chemicals, or silica gel, to treat wall voids.

## SPRUCE BUDWORMS

See: *Leaf Rollers* in this section.

## SQUASH BUGS

*Phylum: Arthropoda*    *Class: Insecta*    *Order: Hemiptera*

The squash bug belongs to the large group of leaf-footed bugs, so-named because both rear legs have leaf-like structures on them. These pests exist throughout the United States and many parts of the world, but are especially common in the southern states. Squash bugs are plant-eaters, and are serious pests of cucurbits (gourds).

### Control

Follow the control procedures given for *Plant Bugs* in this section.

## SQUASH BEETLES

*Phylum: Arthropoda*    *Class: Insecta*    *Order: Coleoptera*

The squash beetle belongs to the ladybird beetle group, and is closely related to the Mexican bean beetle. The squash beetle is pale orange to yellow, with three large spots behind the head, about twelve large spots arranged in two rows on the back, and a large black dot near the tip of the abdomen.

The squash beetle is a serious garden pest.

### Control

Follow the control procedures given for the *Mexican Bean Beetle* in this section.

## SQUIRRELS

*Phylum: Chordata*    *Class: Mammalia*    *Order: Rodentia*

Four types of tree squirrels are common throughout the United States and many parts of the world. These are: the gray, fox, red, and flying squirrels.

Normally these squirrels nest in trees, but occasionally some will occupy attics of houses when conditions are favorable. Squirrels cause varying damage to buildings by gnawing on the structural wood, cables, and electrical wiring (which may cause fires). Squirrels nesting in houses also bring in ectoparasites such as fleas, mites, and lice, which may become established and attack the human inhabitants.

### Control

Frequently squirrels gain access to attics by means of overhanging tree limbs. Cut back limbs that touch or hang close to your home or other buildings. Use sheet metal, mesh wiring, hardware cloth, or other suitable material to seal off the entrances used by squirrels to get inside.

Squirrels inside attics or buildings can be repelled with liberal amounts of naphthalene or PDB (paradichlorobenzene).

To prevent squirrels from climbing trees, install metal bands approximately two feet wide around the tree trunks. These bands, called "baffles," should be placed three to four feet off the ground.

Traps also may be used to reduce the pest-squirrel population on your property. First, however, traps must be installed unset for a few days to allow the squirrels to become accustomed to them.

## STARLINGS

See: *Blackbirds* (Control) and *Pigeons* (Control) in this section.

## STICKTIGHT FLEAS

See: *Fleas* in this section.

## STORED FOOD PESTS

See: *Pantry Pests* in this section.

## STRAW-ITCH MITES

*Phylum: Arthropoda    Class: Arachnida    Order: Acarina*

Straw-itch mites are extremely small. Often called "grain itch" and "straw mattress" mites, they infest straw, stored food, wheat, wood, and straw mattresses inside the home. These mites may attack humans, causing a skin irritation called dermatitis. Human infestation with straw-itch mites usually results from contact with peas, cottonseed, tobacco, broom-corn, beans, straw, hay, grass, and various grains.

### Control

Dispose of all infested materials inside the house. If the mite infestation is extensive, or continues, close off the house tightly, remove all inhabitants, and release three to four aerosol insecticide bombs (or miticide bombs). NOTE: Be sure the bombs are effective against mites before purchasing them.

## STRAWBERRY LEAF ROLLERS

*Phylum: Arthropoda     Class: Insecta     Order: Lepidoptera*

Strawberry leaf rollers are the larvae of moths. The adult moths are small and brownish to grayish in color, frequently with mottled areas or bands on the wings. The front wings are square-tipped. Strawberry leaf rollers attack the foliage of strawberry plants, often doing considerable damage.

### Control

Follow the control procedures given for *Caterpillars* in this section.

## STRAWBERRY WEEVILS

*Phylum: Arthropoda     Class: Insecta     Order: Coleoptera*

Adult strawberry weevils are about one-fourth of an inch long, usually yellowish to brownish in color, with a snout that is about one half as long as the body. These pests may cause serious damage to strawberries.

### Control

Apply a contact spray or dust to affected plants when the weevils are visible. Otherwise, a residual spray or dust should be applied. Repeat this treatment as needed.

The following insecticides are recommended: 1–2% Gardona or rotenone; 2% malathion or Sevin; 0.5% diazinon; or 1% lindane.

## SUGARBEET ROOT MAGGOTS

See: *Root Maggots* in this section.

## SUNFLOWER PESTS

The chemical Supracide is recommended for several sunflower pests. Follow label directions when applying this insecticide.

## SUNSPIDERS

*Phylum: Arthropoda      Class: Arachnida      Order: Solpugida*

Sunspiders (solpugids) occur mostly in dry, tropical, and warm-temperature areas. They resemble true spiders, but have two sharp chelicerae (mouth pincers) that curve *downward* from the mouth. Though fearsome looking, sunspiders are quite harmless. They are mostly active at night, and tend to hide under stones, logs, boards, and other objects during the day. Occasionally sunspiders may invade homes where they become unwelcome pests.

### Control

Follow the control procedures given for *Spiders* in this section.

## SWALLOW BUGS

*Phylum: Arthropoda      Class: Insecta      Order: Hemiptera*

The swallow bug is closely related, and quite similar, to the bedbug. The swallow bug, however, is more rounded in front, and there are long hairs on its body.

These bugs inhabit swallow nests. When these nests are located under eaves or within attics, and are knocked down, swallow bugs often invade the home in large numbers and attack the human inhabitants. Swallow bug bites cause reddening and itching of the skin.

### Control

Remove all swallow nests and other bird nests from eaves and attics. First, however, spray the nests thoroughly with a contact spray *before* tearing them down, if this is possible. The initial spraying should kill all swallow bugs present in the nests, thus preventing them from invading the house.

For swallow bugs inside, close off the house tightly and apply a space spray, or release several aerosol bombs. Allow this treatment to work for several hours before re-entering and venting the house. An alternative inside treatment is to apply a residual spray to cracks, crevices, floors, walls, bedsprings, bed frames, moldings, baseboards, and other hiding areas.

Insecticides recommended are as follows: contact and space sprays: 1–2% DDVP, pyrethrins, SBP-1382®, or resmethrin; residual sprays: 2% malathion; 1% diazinon, 0.5% Dursban, or 1% Baygon or lindane.

## SWEET POTATO WEEVILS

*Phylum: Arthropoda    Class: Insecta    Order: Coleoptera*

The adult sweet potato weevil is a slender, elongated, ant-like insect about one-fourth of an inch long. This weevil is reddish-brown on the front thorax and blue-black on the back. Larvae of this weevil are called sweet potato borers, because they bore into the vines and roots, often killing potato plants. These larvae often continue to bore into the tubers after they are harvested, and adult weevils may emerge from sweet potatoes in storage. These weevils are serious pests, especially in the southern United States.

### Control

Follow label directions and apply to potato plants (soil) infested with the larvae (borers) any of the following insecticides: Temik, Furadan, Syston, Di-Syston, Monitor, Azodrin, Thimet, or pirimicarb.

Adult weevils on the plants may be treated with: Gardona, rotenone, sabadilla, Orthene, Sevin, or malathion.

## TAPESTRY MOTHS

See: *Carpet Moths* in this section.

# TARANTULAS

*Phylum: Arthropoda     Class: Insecta     Order: Araneida*

Tarantulas are the largest of all spiders. Their hairy bodies and large size present a frightening appearance, but actually they are relatively harmless. Although a few South American tarantulas do possess a deadly bite, the venom of North American species generally is no more harmful than a bee sting—unless, of course, one is hypersensitive (highly allergic). In fact, tarantulas rarely bite humans, even when handled roughly. Rather, they tend to remain sluggish and quiet. Some people, incidentally, keep tarantulas as pets.

Tarantulas are nocturnal and hide during the day in ground burrows and cavities. The females live for many years, and usually molt approximately once per year.

### Control

Tarantulas rarely become pests. Occasionally, however, they may enter homes, though rarely in large numbers. Should treatment for tarantulas seem indicated, follow the control procedures given for *Spiders* in this section.

# TARNISHED PLANT BUGS

*Phylum: Arthropoda     Class: Insecta     Order: Hemiptera*

The tarnished plant bug is about one-fourth of an inch long, and varies from pale brown to almost black in color. It is a serious pest of vegetation, feeding on a variety of both wild and cultivated plants and flowers.

### Control

Follow the control procedures given for *Plant Bugs* in this section.

# TENT CATERPILLARS

See: *Caterpillars* in this section.

# TERMITES

*Phylum: Arthropoda     Class: Insecta     Order: Isoptera*

Termites are highly destructive insects that cause the deterioration of housing and wooden structures throughout the United States and many parts of the world. They are colonial (social) insects with a caste system, consisting of workers, soldiers, and reproductives. Workers are the most abundant and do all the damage to wood. Soldiers guard the colony against intrusion by other insects, and the reproductives (males and females with wings) swarm out of the colony each spring and found new colonies.

Termites may be confused with ants which they somewhat resemble. In fact, many people refer to termites as "flying ants." The following differences, however, will enable you to distinguish termites from ants:

1. The termite abdomen joins the thorax with a broad waist, while the ant abdomen joins the thorax with a narrow, threadlike waist.

2. The two pairs of termite wings are of equal length and have many veins present in them. Ant forewings, however, are longer than the hind wings, and both pairs have but few veins.

3. Termite antennae are generally straight and bead-like in appearance. Ant antennae are elbowed.

NOTE: It will be necessary to use a strong hand lens or magnifying glass to distinguish these differences.

Approximately sixty species of termites exist in the United States. These are divided into three main types: (1) *Subterranean Termites* which live within the ground and are the most serious and destructive group; (2) *Drywood Termites* (discussed separately in this section); and (3) *Wetwood Termites* which attack living trees, and are of only minor importance. These main groups are discussed in more detail below.

***Subterranean Termites.*** These insects live within the soil, and construct earthen or mud tunnels from the ground up to the timbers of buildings which they attack. A colony of subterranean termites may contain as many as 250,000 individuals. They can live out of sight—and undetected—for years, regularly moving back and forth through their tunnels from the ground nest below to the structural wood above, which they slowly destroy from the inside. However, the reproductives swarm once per year, usually on a warm spring day and especially after a rain. When swarming occurs, the winged termites may appear by the hundreds, or thousands, as they seek to move out and establish a new colony. NOTE: the appearance of these swarming winged termites almost certainly indicates the presence of a serious termite infestation beneath the house or building.

**Drywood Termites.** See separate entry in this section.

**Wetwood Termites.** These insects attack damp wood and living trees. They are of minor importance to the average person.

### Control

Drywood termite control is discussed separately in the section: *Drywood Termites.*

Wetwood termites are only of minor importance, and control procedures will not be covered here.

Subterranean termite detection and control is very difficult. This requires a trained and properly-equipped specialist, and is simply beyond the capacity of the average person. Moreover, the chemicals used in subterranean termite control are restricted and cannot legally be purchased nor used by the homeowner. It is suggested, therefore, that all readers have their homes inspected and, if necessary, treated, for subterranean termites by professional pest control firms.

## THRIPS ("Oat Bugs")

*Phylum: Arthropoda*     *Class: Insecta*     *Order: Thysanoptera*

Thrips are very small, slender insects, with some species hardly visible to the naked eye. Nymphs are whitish to yellowish or orange, while adults may vary from tan to brown, blackish-brown, or even black. These insects tend to fly, run, or leap when disturbed.

Despite their small size, thrips are important pests of various crops, plants, and flowers. They attack and destroy buds and blossoms, and cause leaves to whiten, curl, or deform. Thrips also will invade homes, especially from nearby fields. They often bite humans, producing a pricking sensation that later itches. Thrips may be brought into the house on blankets, sheets, and other clothing that is hung outside.

### Control

Spray or dip plants infested with thrips. Emulsions (detergent-water-insecticide), oil-insecticide solutions, or suspensions are recommended for best results. The following chemicals are recommended: Baygon, diazinon, Dursban, malathion, DDVP, or lindane.

Repeat this treatment as needed.

# TICKS

See: *American Dog Ticks, Brown Dog Ticks,* and *Soft Ticks* in this section.


# TISSUE PAPER BUGS

See: *Odd Beetles* in this section.


# TOBACCO MOTHS

*Phylum: Arthropoda      Class: Insecta      Order: Lepidoptera*

The tobacco moth is light brownish gray with two light-colored bands extending across each forewing. The hind wings are uniformly gray. The wingspan of this moth is about five-eighths of an inch.

Larvae (worms) of this moth migrate to sheltered locations where they pupate. These moths overwinter and appear again in May or June. The tobacco moth is a pest of stored tobacco leaves, chocolate, cocoa, cereals, beans, coffee, cottonseed, flour, nuts, seeds, spices, and dried fruits.


### Control

Large-scale infestations of this moth in tobacco sheds or warehouses will require fumigation and/or other special control methods applied by licensed pest control firms. Smaller infestations in homes, storage rooms, etc. may be controlled with sprays of: 0.2–0.5% pyrethrins in oil, or 1–2% DDVP, or 1.5% SBP-1382®.


# TOMATO FRUIT WORMS

See: *Corn Earworms* in this section.


# TOMATO HORNWORMS

*Phylum: Arthropoda      Class: Insecta      Order: Lepidoptera*

Tomato hornworms are the larvae (caterpillars) of sphinx or hawk moths. These larvae are large green caterpillars (worms) with a horn-like protrusion on the head. They feed on tomatoes, tobacco, and potato plants, often doing serious damage.

### Control

See: *Caterpillars* (Control) in this section.

## TOMATO PSYLLIDS

See: *Potato Psyllids* in this section.

## TREEHOPPERS

*Phylum: Arthropoda    Class: Insecta    Order: Homoptera*

Treehoppers are recognized by the large shield-like device that covers the head and extends back over the body. This shield often assumes peculiar shapes. Some treehoppers thus appear humpbacked, while others have various spines, keels, horns, and other projections on the foreback. They are one-half of an inch or less in length, and the wings are concealed.

Treehoppers feed mainly on trees and shrubs. Most damage, however, is done by the egg-laying process. The buffalo treehopper, for example, lays its eggs in slits cut in the bark of apple tree twigs and other trees. Often this causes part of the twig to die and drop off. The eggs lie dormant within the twigs and hatch in the spring.

### Control

Large trees and/or orchards will require a special power-operated sprayer or fogger for effective treatment. Smaller trees and shrubs may be treated with a pressurized hand-operated sprayer.

Treat all affected trees and shrubs in the spring when treehoppers first appear. Repeat this treatment as needed, especially later in the season before eggs are laid for overwintering.

Use a residual or contact spray of: 3% malathion or Sevin; 2% diazinon; or 1–2% Orthene, Dimecron, or Cygon.

## VARIED CARPET BEETLES

*Phylum: Arthropoda      Class: Insecta      Order: Coleoptera*

Adult varied carpet beetles are much smaller than the regular black carpet beetle. The varied carpet beetle measures only a fraction of an inch in length, and is marked with various white, brown, orangish, and yellowish spots. A strong hand lens or magnifying glass will reveal the rounded shape and various colored markings.

These pests lay their eggs in various locations, and hatching occurs in about eighteen days. Larvae molt several times over a seven to eleven month period. These larvae are about one-fourth of an inch long with three pairs of tufts on the back. They feed on various animal products such as woolens, feathers, carpets, hides, and other items. They also attack such household items as rye, meal, corn, red pepper, and other foods. They may live in bird nests, on dead animals, and in insect collections.

### Control

For infestations of stored food items, follow the control procedures given for *Pantry Pests* in this section.

For infestations of woolens, carpets, and similar materials, apply the control procedures recommended for *Buffalo Bugs* and/or *Casemaking Moths* in this section.

## VEGETABLE PESTS

Most vegetable pests, except rabbits, belong to the following insect and arachnid groups: aphids, plant lice, ants, caterpillars, beetles, weevils, plant bugs, planthoppers, leafhoppers, thrips, mites, rootworms, and loopers. As with other types of pests, you should first attempt to identify the kind of insect or arachnid that is attacking your vegetable before applying control measures. If, however, you are unable to identify the pest, it is suggested that you apply, in residual spray or dust form, any of the insecticides listed below.

Recommended insecticides for vegetable pests are: rotenone, Gardona, sabadilla, ryania, Sevin, Guthion, malathion, acephate, methoxychlor, Thiodan, or pyrethrins.

## VELVET ANTS

*Phylum: Arthropoda*    *Class: Insecta*    *Order: Hymenoptera*

Velvet ants are large, and are so-named because of their velvety appearance. Males have wings and are larger than the females which are wingless. Colors are usually reddish-orange with one or more broad black stripes across the abdomen.

These large ants may become pests when they swarm in large numbers in yards, turfs, and other areas. Female (wingless) ants can inflict a painful sting.

### Control

Follow the control procedures given for *Ants* in this section.

## VINEGAR FLIES

See: *Pomace Flies* in this section.

## VINEGARROONS

See: *Whip Scorpions* in this section.

## WALKING STICKS

*Phylum: Arthropoda*    *Class: Insecta*    *Order: Orthoptera*

Walking sticks are mainly tropical insects, but exist widely in other areas as well, especially in the southern United States. These are long, slender insects that greatly resemble twigs. Some tropical species, however, resemble leaves. Walking sticks are slow-moving plant-eaters that are related to grasshoppers and mantids. They live on trees and shrubs. Walking sticks can emit a foul-smelling substance from glands on the thorax.

Normally walking sticks do not exist in numbers large enough to do serious damage to trees, shrubs, or plants in the United States. Occasionally, however, they may indeed reach numbers large enough to damage vegetation. Eggs are laid on the ground and hatch only in alternate years. Thus, walking sticks tend to appear in alternate years.

The young are greenish in color, while adults are brown, much like the twigs they resemble. Walking sticks in the United States vary in length from an inch or so up to six or seven inches. Some tropical species, however, may reach a foot or more in length. Walking sticks do not bite humans.

### Control

When walking sticks are present in large numbers on trees, shrubs, or plants, treat these with a contact spray. Depending upon the size of the trees and shrubs involved, a power-operated sprayer may be required. Repeat this treatment as needed.

The following insecticides are recommended: 3% malathion, Sevin, Orthene, Lannate, or toxaphene; 1.5% Baygon, diazinon, or lindane; or 1% Dursban.

## WASPS

*Phylum: Arthropoda     Class: Insecta     Order: Hymenoptera*

Wasps of wide variety exist throughout many parts of the world. Some of the most common and dangerous wasps belong to the *Vespidae*, or "paper making" group. These wasps build nests of paper-like material that contain cells. A queen places an egg in each cell. After the eggs hatch, the queen brings food to the larvae that are developing within the nest cells until they pupate.

Following pupation, the adult wasps are, anatomically speaking, females; but at this time, their reproductive organs do not develop or become functional. Thus, these adults become workers and take over the nest building and caring for the young from the queen. Near the end of summer, however, the queen lays a number of eggs that develop into males and young queens. Mature males and females emerge, leave the nest, and mate. The nest is thus left abandoned.

Paper wasps are of medium size, and vary in color from black and yellow to brownish and reddish. These wasps can and do sting humans. Like bees, the venom of wasps is toxic and can be dangerous to sensitive persons.

**Control**

Control of paper wasps should begin by removing nests which usually are visible on the outer walls of houses and outbuildings, or on trees, shrubs, fence posts, etc. Before knocking down these nests, however, you should first spray them thoroughly from a safe distance, using a pressurized sprayer or a special "wasp/hornet" dispenser can that squirts a stream of insecticide ten to fifteen feet.

Wasps that swarm around your home, especially on shrubs and plants, should be treated with a contact or residual spray.

The following insecticides are recommended: Sevin, Baygon, resmethrin, diazinon, malathion, Dursban, Ficam, DDVP, pyrethrins, Wasp Freeze, SBP-1382®, Gardona, and thanite.

## WATERBUGS

See: *Cockroaches* (*American and Oriental*) in this section.

## WAX MOTHS

*Phylum: Arthropoda*     *Class: Insecta*     *Order: Lepidoptera*

Wax moths often infest the combs of honeybees in wall voids, attics, chimneys, and other hidden areas of the house. These moths and their worm-like larvae may then invade the living area of the house where they become annoying pests.

Adult wax moths are gray or pale brown, with black markings and "bumps" on the forewings. They are about one half of an inch to slightly less than one inch in length. Larvae vary from white to yellow, with black or brown heads. In some cases, the backs of these larvae may be nearly black.

**Control**

When wax moths and/or their larvae exist inside the home, look for honeybee combs in wall voids, attics, chimneys, and other such hidden areas. Either remove these combs or treat them thoroughly with a spray or dust.

Use a contact spray to kill visible moths and larvae in the living area of the home.

The following insecticides are recommended: 1–2% DDVP or SBP-1382®; 2% pyrethrins; 2–3% methoxychlor; 0.5% Ficam; 2% malathion; or 0.5% lindane.

## WEBBING CLOTHES MOTHS

*Phylum: Arthropoda*     *Class: Insecta*     *Order: Lepidoptera*

The webbing clothes moth is the most common, and most destructive, clothes moth in the United States. These moths tend to damage clothing in dark, hidden areas—especially under collars and cuffs. The larvae (worms) of this moth are quite active, and may be seen crawling on clothes, on the floor, or beneath infested furniture. These larvae may be found feeding between carpets and the floor, leaving a webbing pattern as they go. They also feed on rugs, upholstered furniture, furs, wool, piano felts, and similar material.

Adult webbing clothes moths have uniformly golden colored bodies and wings, with a few reddish-golden hairs on top of the head. The wingspan of this moth is slightly less than one-half of an inch. Fully developed larvae are a shiny creamy white in color, and are about one half of an inch long.

### Control

Follow the control procedures given for *Casemaking Moths* and *Buffalo Bugs* in this section.

## WEBWORMS

*Phylum: Arthropoda*     *Class: Insecta*     *Order: Lepidoptera*

Webworms are the larvae of moths. These worm-like larvae form tubular webs in the ground that may go two feet deep. When disturbed they retreat into these underground webs. These larvae feed on the roots of various grasses and plants, and often form webs on the grass blades near the ground. Webworms often completely destroy young corn plants, and are serious pests in yards, lawns, gardens, and turfs.

### Control

Follow label directions and apply to infested sod, yards, lawns, turfs, plants, etc. any of the following insecticides: Sevin, diazinon, trichlorfos, Dursban, Aspon, ethoprop, chlormephos, or Mocap.

# WEEVILS

*Phylum: Arthropoda    Class: Insecta    Order: Coleoptera*

Weevils belong to the very largest order of insects, the beetles. Weevils thus are a form of beetle, but differ somewhat in appearance. Typically, weevils have snouts which vary from long and slender to short, wide, or blunt. Many weevils are serious pests of various grains, stored food, peas, beans, and various plants and crops.

Many important weevils already have been discussed in this handbook. Three other important types are discussed below.

**Black Vine Weevil.** This weevil is about one-third of an inch long, brownish to black in color, and with small patches of golden scales scattered over the back. Adults are nocturnal and feed on various plants. The larvae are white, about one-third of an inch long, and live in the soil where they feed on, and damage, the roots of plants. Black vine weevils often live on hedges and other dooryard plants, which they may defoliate almost completely. These weevils may also migrate into the house. Although they neither bite nor do any damage, they nevertheless are annoying pests when they invade the home.

**Alfalfa Weevil.** This weevil is a natural pest of clover and alfalfa. It lives in the western United States, especially in Arizona and California. The adult is about one-fifth of an inch long, grayish brown to almost black in color, and has short grayish hairs over the body, as well as a long, slender snout. These weevils frequently migrate from clover or alfalfa fields to nearby trees, fence rows, hedges, and buildings. They may invade homes in large numbers.

**Strawberry Root Weevil.** This weevil is about one-fourth of an inch long and blackish-brown in color. The larvae feed on the roots of wild and cultivated strawberry plants, brambles, and evergreen trees. Root weevils may invade the home in large numbers, especially in the spring and fall.

### Control

Control measures given here pertain only to weevils in and around the home, *not* to large fields of clover, alfalfa, strawberries, or other plants.

Weevils originate outdoors, thus control must begin outside. Use a residual spray or dust to treat all infested plants, trees, hedges, flowers, etc. Use a residual spray to treat the ground around your home, as well as the foundation, lower walls, crawl spaces, windows, doors, and steps.

When weevils invade the house in large numbers, vacuuming them up is the quickest and most effective control method.

The following insecticides are recommended: 1.5% Baygon; 1% lindane; 1% diazinon; 0.5% Dursban; 2% malathion or Sevin; or 0.5% Ficam.

## WEED BUGS

*Phylum: Arthropoda      Class: Insecta      Order: Hemiptera*

The adult weed bug is uniformly brown and one-third of an inch or more in length. It is somewhat similar to the grass bug which was discussed earlier; but, instead of grass, the weed bug feeds on weeds. This insect is a house pest in some areas of the United States, particularly in California. These bugs often invade homes in large numbers, especially in areas of new subdivisions, and around fields, lots, ditch banks, and levees.

Weed bugs, which emit a disagreeable odor, enter houses by crawling beneath doors and windows. Inside, they stain draperies, furniture, and rugs with their excrement.

### Control

Use a residual spray to treat the margins of fields, levees, ditch banks, and similar weed-infested areas adjacent to your home. Also treat the ground around the house, the foundation, lower walls, crawl spaces, windows, steps, and doors with the same residual spray.

Bugs that are visible inside should be vacuumed up, if possible. Following this, the entire house should be closed off tightly and several aerosol insecticidal bombs set off. Allow these to work for several hours before re-entering and venting the house. NOTE: Purchase only those aerosol bombs that are listed for bugs, or that contain any of the insecticides listed below.

Recommended insecticides are: Sevin, malathion, diazinon, Baygon, Dursban, lindane, and Ficam.

## WESTERN PINE BEETLES

*Phylum: Arthropoda      Class: Insecta      Order: Coleoptera*

The western pine beetle is an important pest of ponderosa and Coulter pines in western North America. The adult beetle is dark brown to black in color, and about one-fifth of an inch long. Adults, eggs, and larvae occur in, or beneath, the bark of pines. Adult beetles attack pines in the spring, and one to three generations are produced each year.

### Control

Thoroughly treat tree trunks and bark with a *residual* spray of: 1% lindane, Dursban, or Baygon. Repeat this treatment as needed.

## WHARF BORERS

*Phylum: Arthropoda     Class: Insecta     Order: Coleoptera*

The wharf borer is a woodboring beetle. Adults are narrow, slightly less than one half of an inch long, and brownish to reddish-yellow above. The eyes, sides of the thorax, legs, and lower body are somewhat blackish, and the entire body is covered with dense yellow hairs. The antennae of this beetle are unusually long. The larvae are about one and one half inches long, cream colored, and covered with brown hairs.

The wharf borer is a serious pest of wood and timbers throughout many parts of the world. It is apt to attack wood pilings beneath buildings, foundation timbers, structural timbers in cellars and basements, and even wood floors.

### Control

Infested wood timbers and other structures should be drilled and injected with a 5% pentachlorophenol-in-oil solution, Vikane, or 1% lindane solution. Drill small holes that reach the beetle galleries or tunnels within the wood, and use a pressurized sprayer to force the insecticide throughout the tunnels within the wood.

Sevin and Thiodan also seem to be effective in controlling this beetle when used as a wettable powder in water at 3–4% strength.

## WHARF RATS

See: *Rats (Norway)* in this section.

## WHEEL BUGS

*Phylum: Arthropoda*      *Class: Insecta*      *Order: Hemiptera*

Wheel bugs are large grayish bugs, so named because a semicircular ridge that resembles half of a cogwheel sits atop the thorax. Knobs on the head, the cogwheel structure, and the eyes are black, while the legs and antennae are tinged chestnut reddish. These bugs often reach one and one-fourth inches in length. The wheel bug lives in the eastern and southern United States, and may be found as far west as New Mexico. Eggs of this bug are glued to twigs, tree bark, fence rails, or beneath the eaves of houses and buildings.

The bite of the wheel bug is very painful, often producing inflammation and swelling that can last for several days. Several weeks may be required for the bite to heal completely.

### Control

To control wheel bugs that live on the ground, trees, shrubs, plants, etc., apply a residual spray to all infested surfaces.

Normally wheel bugs do not invade houses in large numbers. Should this occur, however, use a contact spray to kill those present. If the infestation continues, apply a residual spray to all affected areas of the house.

The following insecticides are recommended: 1.5% Baygon; 3% Sevin or malathion; 0.5% Dursban; 1% diazinon or lindane; or 2% trichlorfon.

## WHIP SCORPIONS (Vinegarroons)

*Phylum: Arthropoda*      *Class: Arachnida*      *Order: Pedipalpida*

The whip scorpion is a fearsome looking arachnid (not insect) that is actually harmless to man. Whip scorpions resemble both spiders and true scorpions to some degree. There are two body parts, with a slender whip-like tail attached to the end of the abdomen. There are eight eyes and a pair of large pincers (pedipalps) used to seize and tear apart prey. Whip scorpions, when disturbed, discharge a stream of fluid that smells like vinegar from glands on each side of the anus, hence the name "vinegarroon." Whip scorpions vary in size from a mere fraction of an inch up to nearly three inches in length.

These arachnids are nocturnal, and tend to hide during the day beneath leaves, rocks, logs, boards, and in other dark, moist areas. Occasionally they may invade homes where they become annoying, yet harmless, pests.

### Control

Follow the control procedures given for *Scorpions* in this section when whip scorpions occur outdoors. When they occur indoors, apply the control procedures given for *Spiders*.

## WHITE ANTS

See: *Termites* in this section.

## WHITEFLIES

*Phylum: Arthropoda*    *Class: Insecta*    *Order: Homoptera*

Whiteflies are very small insects, a mere fraction of an inch long, that resemble tiny moths. Wings of both sexes are covered with a waxy powder or white dust. During development from eggs to adults, whiteflies at one stage resemble scales, for which they may be mistaken. Whiteflies are most common in tropical and sub-tropical parts of the world, but a few species exist in the United States.

The North American species are serious pests of citrus trees and greenhouse plants. Damage results from whiteflies sucking the sap and juices from the leaves. These insects also secrete honeydew as they feed, which often causes the growth of a sooty fungus that further damages the plants.

### Control

When infested plants are small and/or few in number, dip the leaves and stems in an emulsion (detergent-water-insecticide), a suspension, or an oil solution of insecticide. These particular forms of insecticide are necessary because ordinary water-based sprays are not effective against the waxy covering found on whiteflies and scales.

Larger trees, shrubs, plants, and/or large numbers of plants will require treatment with a pressurized sprayer using one of the insecticide formulations given above. NOTE: It is especially important to treat the underside of leaves where whiteflies and aphids are most commonly found.

The following insecticides are recommended: malathion, rotenone, lindane, Dursban, acephate, demeton, diazinon, Trithion, Guthion, nicotine sulfate, or kinoprene (Enstar). Follow label directions when applying any of these chemicals.

## WHITE OAK LEAF MINERS

See: *Leaf Blotch Miners* in this section.

## WHITE PINE WEEVIL

*Phylum: Arthropoda    Class: Insecta    Order: Coleoptera*

The white pine weevil belongs to a larger group of pine weevils that usually are brownish in color, cylindrically shaped, and measure about one-third of an inch in length. Larvae of these weevils tunnel in the terminal leaders of trees and kill them. This, in turn, causes a bend to form partway up the trunk. This weevil is a specific pest of white pine.

### Control

Thoroughly treat tree trunks, bark, leaves, and stems with a residual spray of: 1% lindane, Baygon, or Dursban.

## WIGGLERS

*Phylum: Arthropoda    Class: Insecta    Order: Diptera*

Wigglers are the larvae of mosquitoes, so named because of their very apparent wiggling motion in water. Around the home, wigglers occur in standing water containers such as rainbarrels, buckets, cans, clogged gutters, fish ponds, bird baths, and tires. If not destroyed, wigglers will in a few days develop into flying adult mosquitoes that are annoying and dangerous pests.

**Control**

Pour out standing water from all containers, if possible. This quickly destroys all wigglers present because they cannot survive out of water. If all water containers cannot be emptied, however, apply any of the following chemicals to the water surface of the container: Abate, Altosid, diesel fuel, kerosene, or motor oil.

## WIREWORMS

*Phylum: Arthropoda*    *Class: Insecta*    *Order: Coleoptera*

Wireworms are the larvae of click beetles. The adult beetles have elongated, parallel-sided bodies that are rounded at both ends. Most of these beetles are between one half of an inch to one and one half inches in length. They produce a peculiar sharp "clicking" sound, and are able to jump. The larvae, or wireworms, often are very destructive pests of newly planted seed and the roots of corn, beans, potatoes, cabbage, and cereal plants.

**Control**

Follow label directions and apply to soil infested with wireworms any of the following insecticides: ethoprop (Mocap), aldicarb, Aspon, Ficam (bendiocarb), chlormephos, Dasanit (fensulfothion), or terbufos (Counter).

## WOODBORING INSECTS

See: *Ants (Carpenter), Bees (Carpenter), Old House Borers, Powder-Post Beetles, Drywood Termites, Termites (Subterranean), Powder-Post Termites, Woodwasps,* and *Wharf Borers.*

## WOOD DECAY

Wood decay is caused by fungus growth which, in turn, results from dampness and moisture. Thus, the first step in controlling wood decay is to eliminate all moisture and dampness. Secondly, if the wood is still sound (not rotten) allow it to dry completely, then treat it with heavy, thick coats of PCP (pentachlorophenol) in oil, or treat with creosote. This solution should be applied in two or three separate applications, each of which must be brushed on heavily. NOTE: These chemicals will not be effective in preventing wood decay unless moisture and dampness are eliminated.

## WOODPECKERS

*Phylum: Chordata      Class: Aves      Order: Piciformes*

Woodpeckers sometimes are bothersome pests when they damage structural wood and disturb humans with their persistent pecking or drumming on houses or buildings. These birds are apt to attack wooden shingles, eaves, posts, and other wooden structures on buildings. Male woodpeckers drum or peck for courtship purposes and to establish their individual territories, rarely to find insects, despite this common belief. In most cases a woodpecker pest problem results from the activity of one bird alone.

### Control

Hang strips of tin foil or aluminum foil approximately three inches wide by about three feet long from eaves and other areas attacked by woodpeckers. This simple technique is effective in many cases.

An alternative repellent method calls for treating wood with a 10% PCP (pentachlorophenol)-in-oil solution. First, however, this solution should be tested on a small section of the wood to be treated to see whether staining occurs.

Finally, the woodpecker pest may be trapped by securing a wooden-base rat trap, baited with suet, to wooden areas under attack.

## WOODWASPS (Horntails)

*Phylum: Arthropoda      Class: Insecta      Order: Hymenoptera*

Woodwasps commonly are called "horntails" because of the long hornlike projection that extends from the end of the abdomen. In some species this "horntail" may exceed the length of the body itself. Wood-wasps are rather large insects, usually about one inch long, with the females being much larger than the males. Adults usually are black or metallic blue in color. Some, however, are colored with combinations of red, black, and yellow.

Woodwasps produce a conspicuous buzzing noise as they actively fly about in bright sunshine. These insects are especially attracted to coniferous trees that are weakened or dying from disease, fire, previous insect attack, or other causes. Woodwasps also attack wooden walls and the structural timbers of houses and buildings. The female uses her long ovipositer (egg-depositor) to deposit eggs deeply (one and three-fourths inches or more) into wood. The eggs hatch in three to four weeks, and the resulting larvae burrow into the wood at right angles to the egg channel.

Larvae of woodwasps are worm-like, cylindrical, S-shaped, and whitish, creamy, or yellow in color. Following pupation within the wood, the adult woodwasps chew their way to the surface and emerge, leaving a perfectly round exit hole. These holes in most cases cannot be distinguished from those made by some woodboring beetles. The entire woodwasp life cycle requires at least two years, but may be twice this long in some cases.

### Control

For wooden walls infested with woodwasps, drill holes into the plaster. Force large quantities of an insecticidal dust into these holes under pressure. NOTE: a modified water-type fire extinguisher can be used for this treatment method. When treatment is complete, plug the holes with matching plaster or caulk.

Recommended insecticides are: Drie-die (or another form of silica aerogel); 1% Baygon, lindane, or diazinon dust; 0.5% Dursban dust; or 2% malathion dust.

## WOOLLYBEARS

See: *Caterpillars* in this section.

## YELLOWJACKETS

*Phylum: Arthropoda*    *Class: Insecta*    *Order: Hymenoptera*

Yellowjackets are paper wasps that usually nest underground. They also may nest within the wall voids of building foundations. Most yellowjackets are about one half of an inch long, making them the smallest of the paper wasps. Yellowjackets are darkish to black, with yellow bands across the abdomen, and two longitudinal yellow stripes on the thorax.

A single yellowjacket colony may produce up to 10,000 workers during one season. These workers are somewhat unpredictable toward humans who approach the nest. In some cases, they ignore the human intruder; while in other instances they launch a massive attack and sting the person severely. NOTE: Yellowjacket venom is considered by some experts to be the *most dangerous* venom of all wasps and bees in North America.

### Control

NOTE: Wear protective covering or clothing as described in the section on *Hornets*.

A special insecticide product is recommended for the control of yellowjackets: Wasp Freeze. This product comes in a special pressurized can containing highly volatile solvents and pyrethrins for quick knockdown of yellowjackets and other wasps.

As you approach the yellowjacket nest, spray Wasp Freeze into the air with a wide sweeping motion to knock down any flying workers that approach. This method usually clears the area of yellowjackets and, thus, makes it safe to approach the nest itself. Quickly direct the sprayer into the nest opening and empty it without stopping. This action kills those yellowjackets still in the nest, usually resulting in the complete elimination of the colony.

# APPENDIX

# TOXICITY COMPARISON TABLES

The following tables are provided for your convenience in making quick-reference toxicity checks of various pesticides. They are based on the $LD_{50}$ table given earlier and simply classify pesticides alphabetically by name in one of four toxicity categories: 1) Extremely/highly toxic; 2) Moderately toxic; 3) Low toxicity; and 4) Non-toxic.

### Extremely/Highly Toxic Pesticides

aldicarb
Amaze
arsenic trioxide
arsenic
arsenic compounds
Azadrin
Baygon
Bidrin
Bomyl
calcium cyanide
carbofuran
carbon tetrachloride
chlordimeform
chlormephos
Compound 4072
Counter
cyanide
Dasanit

Delnav
dieldrin
Dimecron
dimetilan
dioxathion
Diphacin
disulfoton
Di-Syston
Endrin
EPN (phenylphosphorothioate)
ethion
famphur
fluoroacetamide (1081®)
Furadan
Guthion
hydrogen cyanide
Kepone
Meta-Systox-R

mevinphos
Mocap
Monitor
Morocide
Nemafos
Ornitrol
parathion
phorate
Phosdrin
phosphorus
PID®
sodium fluoride
sodium fluoroacetate (1080®)

strychnine
Supracide
Systox
Temik
TEPP (tetra ethyl pyrophosphate)
terbufos
Thimet
Trithion
Vydate
Warbex
Zectran
zinc phosphide

## Moderately Toxic Pesticides

aldrin
ANTU® (alpha-naphthylthiourea)
Avitrol
barium carbonate
Baytex
BHC (benzene hexachloride)
Black Leaf 40
Bolstar
bromophos
BUX®
carbon disulphide
Ciodrin
chlordane
chlordecone
chlorpyrifos
Cygon
Cyhexatin
demeton
DDT (dichloro-diphenyl-trichloroethane)
DDVP
diazinon
Dibrom
dichlorvos
dimethoate
Dipterex
DRC-1339®
Dursban

endosulfan
Entex
ethylene dibromide
ethylene dichloride
ethylene oxide
fenthion
Fumarin
heptachlor
Imidan
lindane
Mesurol
methiocarb
methidathion
methomyl
methyl bromide
naled
nicotine sulfate
Paris Green
PCP (pentachlorophenol)
Perthane
phosalone
Phostoxin
Pirimor
Pival
Plictran
Prolate
propoxur

rotenone
Spectracide
starlicide
sulfuryl fluoride
Supracide
Thiodan

toxaphene
trichlorfon
Vapona
Vikane
Zinophos
Zytron

### Low Toxicity Pesticides

allethrin
Ambush
Aramite
Aspon
borax
carbaryl (Sevin)
chlorobenzilate
coumaphos
DDD (see TDE)
D-Con
dimethrin
dinocap
Enstar
fenson
Galecron
Gardona
Genite
Kelthane
Korlan
Lethane
malathion
mirex
Mitox
Morestan
mothballs (flakes)
norbormide
Omite
Orthene

ovex
Ovotran
PDP (paradichlorobenzene)
Pentac
PMP
Pounce
Pydrin
pyrethrum
Rabon
Red Squill
resmethrin
ronnel
Rothane
RoZol
Ruelene
ryania
sabadilla
SBP-1382®
Sevin
TDE
Tedion
tetradifon
thanite
Trolene
Valone
Vendex
warfarin

### Essentially Non-Toxic Pesticides

Abate
Acaralate

Acarol
chlorbenside

methoxychlor                    temephos (Abate)
Perthane                        tetrasol
silica aerogel

# DILUTION TABLES
# FOR MIXING
# PESTICIDES

## LIQUID PESTICIDES

The following mixing and dilution table is provided for your ease and convenience in mixing and diluting liquid pesticides. Most pesticides will come in one of the following concentrations: 100%, 75%, 50%, 35%, 25%, or 20%. The table that follows below is based on these concentrations.

1. First, read your pesticide container label and determine what concentration it is.
2. Determine what strength is recommended to kill the pest that you wish to eliminate.
3. Now estimate the amount of pesticide that you will need to complete the treatment job.

Example: Let's say that your pesticide is a 35% concentrate. And let's say that you wish to treat your kitchen and bathrooms for cockroaches which, in this particular case, happens to call for a 3% pesticide strength. For an average kitchen and bathroom job, two or three *pints* should be enough. Now turn to the dilution table for "35% Concentrate"; look down the left-hand column for 3% . . . 3% is *not* given, as you will see. What to do? Simply use the next *highest* strength—never the next lowest.

The next highest concentration is 5%. Now look down the second column: "Amount Finished Product Desired (oz.)" You want 3 pints, remember? Okay, since 3 pints = 48 ounces, the closest number of ounces is 50. Now go to the third column: "Oz.

Concentrate to Add." In a straight line across from "5% . . . . 50 ounces . . . ." you will see *8.3 ounces*, which is the amount of 35% concentrate to add to 41.7 ounces of diluent (column 4). If your solvent or diluent is water, then you simply add 8.3 ounces of 35% concentrate to 41.7 ounces of water which gives you a 5% strength to apply. The same is true for solvents other than water.

At this point it should be emphasized that, in general, the concentrations of pesticides recommended to kill pests are *rough estimates* only. That is, it is unrealistic to say that a 2.5% solution of pesticide will be *ineffective* where a 3% strength is called for. The critical factor here is to *get as close to the recommended strength as posible* . . . always on the high side—never on the low side of the recommended strength.

## 100% Concentrate

| Strength Desired | Amount Finished Product Desired (Oz.) | Oz. Con. to Add | Oz. Diluent to Add |
|---|---|---|---|
| 30% | 100 (3.1 qt.) | 30 | 70 |
| " | 75 (2.3 qt.) | 22.5 | 52.5 |
| " | 50 (1.6 qt.) | 15 | 35 |
| " | 25 (1.6 pt.) | 7.5 | 17.5 |
| " | 16 (1 pt.) | 4.8 | 11.2 |
| 20% | 100 | 20 | 80 |
| " | 75 | 15 | 60 |
| " | 50 | 10 | 40 |
| " | 25 | 5 | 20 |
| " | 16 | 3.2 | 12.8 |
| 15% | 100 | 15 | 85 |
| " | 75 | 11.3 | 63.7 |
| " | 50 | 7.5 | 42.5 |
| " | 25 | 3.8 | 21.2 |
| " | 16 | 2.4 | 13.6 |
| 10% | 100 | 10 | 90 |
| " | 75 | 7.5 | 57.5 |
| " | 50 | 5 | 45 |
| " | 25 | 2.5 | 22.5 |
| " | 16 | 1.6 | 14.4 |
| 7.5% | 100 | 7.5 | 92.5 |
| " | 75 | 5.6 | 69.4 |
| " | 50 | 3.8 | 46.2 |
| " | 25 | 1.9 | 23.1 |
| " | 16 | 1.2 | 14.8 |

| Strength Desired | Amount Finished Product Desired (Oz.) | Oz. Con. to Add | Oz. Diluent to Add |
|---|---|---|---|
| 6% | 100 | 6 | 94 |
| " | 75 | 4.5 | 70.5 |
| " | 50 | 3 | 47 |
| " | 25 | 1.5 | 23.5 |
| " | 16 | 1 | 15 |
| 5% | 100 | 5 | 95 |
| " | 75 | 3.8 | 71.2 |
| " | 50 | 2.5 | 47.5 |
| " | 25 | 1.3 | 23.7 |
| " | 16 | 0.8 | 15.2 |
| 2.5% | 100 | 2.5 | 97.5 |
| " | 75 | 1.9 | 73.1 |
| " | 50 | 1.3 | 48.7 |
| " | 25 | 0.6 | 24.4 |
| " | 16 | 0.4 | 15.6 |
| 2% | 100 | 2 | 98 |
| " | 75 | 1.5 | 73.5 |
| " | 50 | 1 | 49 |
| " | 25 | 0.5 | 24.5 |
| " | 16 | 0.3 | 15.7 |
| 1% | 100 | 1 | 99 |
| " | 75 | 0.8 | 74.2 |
| " | 50 | 0.5 | 49.5 |
| " | 25 | 0.3 | 24.7 |
| " | 16 | 0.2 | 15.8 |
| 0.5% | 100 | 0.5 | 99.5 |
| " | 75 | 0.4 | 74.6 |
| " | 50 | 0.3 | 49.7 |
| " (approx.) | 25 (approx.) | 0.13 | 24 |
| " (approx.) | 16 (approx.) | 0.1 | 15.9 |

## 75% Concentrate

| Strength Desired | Amount Finished Product Desired (Oz.) | Oz. Con. to Add | Oz. Diluent to Add |
|---|---|---|---|
| 30% | 100 (3.1 qt.) | 37.5 | 62.5 |
| " | 75 (2.3 qt.) | 28.1 | 46.9 |
| " | 50 (1.6 qt.) | 18.8 | 31.2 |
| " | 25 (1.6 pt.) | 9.3 | 15.7 |
| " | 16 (1 pt.) | 6 | 10 |

| Strength Desired | Amount Finished Product Desired (Oz.) | Oz. Con. to Add | Oz. Diluent to Add |
|---|---|---|---|
| 20% | 100 | 25 | 75 |
| " | 75 | 18.8 | 56.2 |
| " | 50 | 12.5 | 37.5 |
| " | 25 | 6.3 | 18.7 |
| " | 16 | 4 | 12 |
| | | | |
| 15% | 100 | 18.8 | 81.2 |
| " | 75 | 14.1 | 60.9 |
| " | 50 | 9.3 | 40.7 |
| " | 25 | 4.8 | 20.2 |
| " | 16 | 3 | 13 |
| | | | |
| 10% | 100 | 12.5 | 87.5 |
| " | 75 | 9.4 | 65.6 |
| " | 50 | 6.3 | 43.7 |
| " | 25 | 3.1 | 21.9 |
| " | 16 | 2 | 14 |
| | | | |
| 7.5% | 100 | 9.4 | 90.6 |
| " | 75 | 7 | 68 |
| " | 50 | 4.8 | 45.2 |
| " | 25 | 2.4 | 22.6 |
| " | 16 | 1.5 | 14.5 |
| | | | |
| 6% | 100 | 7.5 | 92.5 |
| " | 75 | 5.6 | 69.4 |
| " | 50 | 3.8 | 46.2 |
| " | 25 | 1.9 | 23.1 |
| " | 16 | 1.3 | 14.7 |
| | | | |
| 5% | 100 | 6.3 | 93.7 |
| " | 75 | 4.8 | 70.2 |
| " | 50 | 3.1 | 46.9 |
| " | 25 | 1.6 | 23.4 |
| " | 16 | 1 | 15 |
| | | | |
| 2.5% | 100 | 3.1 | 96.9 |
| " | 75 | 2.4 | 72.6 |
| " | 50 | 1.6 | 48.4 |
| " | 25 | 0.8 | 24.2 |
| " | 16 | 0.5 | 15.5 |
| | | | |
| 2% | 100 | 2.5 | 97.5 |
| " | 75 | 1.9 | 73.1 |
| " | 50 | 1.3 | 48.7 |
| " | 25 | 0.6 | 24.4 |
| " | 16 | 0.3 | 15.7 |

| Strength Desired | Amount Finished Product Desired (Oz.) | Oz. Con. to Add | Oz. Diluent to Add |
|---|---|---|---|
| 1% | 100 | 1.3 | 98.7 |
| " | 75 | 1 | 74 |
| " | 50 | 0.6 | 49.4 |
| " (approx.) | 25 | 0.3 | 24.7 |
| " (approx.) | 16 | 0.2 | 15.8 |
| 0.5% | 100 | 0.6 | 99.4 |
| " | 75 | 0.5 | 74.5 |
| " (approx.) | 50 | 0.3 | 49.7 |
| " (approx.) | 25 | 0.1 | 24 |
| " (approx.) | 16 | 0.1 | 15.9 |

## 50% Concentrate

| Strength Desired | Amount Finished Product Desired (Oz.) | Oz. Con. to Add | Oz. Diluent to Add |
|---|---|---|---|
| 30% | 100 (3.1 qt.) | 60 | 40 |
| " | 75 (2.3 qt.) | 45 | 30 |
| " | 50 (1.6 qt.) | 30 | 20 |
| " | 25 (1.6 pt.) | 15 | 10 |
| " | 16 (1 pt.) | 9.6 | 6.4 |
| 20% | 100 | 40 | 60 |
| " | 75 | 30 | 45 |
| " | 50 | 20 | 30 |
| " | 25 | 10 | 15 |
| " | 16 | 6.4 | 9.6 |
| 15% | 100 | 30 | 70 |
| " | 75 | 22.6 | 52.4 |
| " | 50 | 15 | 35 |
| " | 25 | 7.6 | 17.4 |
| " | 16 | 4.8 | 11.2 |
| 10% | 100 | 20 | 80 |
| " | 75 | 15 | 60 |
| " | 50 | 10 | 40 |
| " | 25 | 5 | 20 |
| " | 16 | 3.2 | 12.8 |
| 7.5% | 100 | 15 | 85 |
| " | 75 | 11.2 | 63.8 |
| " | 50 | 7.6 | 42.4 |

| Strength Desired | Amount Finished Product Desired (Oz.) | Oz. Con. to Add | Oz. Diluent to Add |
|---|---|---|---|
| 7.5% | 25 | 3.8 | 21.2 |
| " | 16 | 2.4 | 13.6 |
| 6% | 100 | 12 | 88 |
| " | 75 | 9 | 66 |
| " | 50 | 6 | 44 |
| " | 25 | 3 | 22 |
| " | 16 | 2 | 14 |
| 5% | 100 | 10 | 90 |
| " | 75 | 7.6 | 67.4 |
| " | 50 | 5 | 45 |
| " | 25 | 2.6 | 22.4 |
| " | 16 | 1.6 | 14.4 |
| 2.5% | 100 | 5 | 95 |
| " | 75 | 3.8 | 71.2 |
| " | 50 | 2.6 | 47.4 |
| " | 25 | 1.2 | 23.8 |
| " | 16 | 0.8 | 15.2 |
| 2% | 100 | 4 | 96 |
| " | 75 | 3 | 72 |
| " | 50 | 2 | 48 |
| " | 25 | 1 | 24 |
| " | 16 | 0.6 | 15.4 |
| 1% | 100 | 2 | 98 |
| " | 75 | 1.6 | 73.4 |
| " | 50 | 1 | 49 |
| " | 25 | 0.6 | 24.4 |
| " | 16 | 0.4 | 15.6 |
| 0.5% | 100 | 1 | 99 |
| " | 75 | 0.8 | 74.2 |
| " | 50 | 0.6 | 49.4 |
| " (approx.) | 25 (approx.) | 0.3 | 24.7 |
| " (approx.) | 16 (approx.) | 0.2 | 15.8 |

## 35% Concentrate

| Strength Desired | Amount Finished Product Desired (Oz.) | Oz. Con. to Add | Oz. Diluent to Add |
|---|---|---|---|
| 30% | 100 (3.1 qt.) | 87 | 13 |
| " | 75 (2.3 qt.) | 65.3 | 9.7 |

| Strength Desired | Amount Finished Product Desired (Oz.) | Oz. Con. to Add | Oz. Diluent to Add |
|---|---|---|---|
| " | 50 (1.6 qt.) | 43.5 | 6.5 |
| " | 25 (1.6 pt.) | 21.8 | 3.2 |
| " | 16 (1 pt.) | 13.9 | 2.1 |
| 20% | 100 | 58 | 42 |
| " | 75 | 43.5 | 31.5 |
| " | 50 | 29 | 21 |
| " | 25 | 14.5 | 10.5 |
| " | 16 | 9.3 | 6.7 |
| 15% | 100 | 43.5 | 56.5 |
| " | 75 | 32.8 | 42.2 |
| " | 50 | 21.8 | 28.2 |
| " | 25 | 11 | 14 |
| " | 16 | 7 | 9 |
| 10% | 100 | 14.5 | 85.5 |
| " | 75 | 11 | 64 |
| " | 50 | 7.3 | 42.7 |
| " | 25 | 3.8 | 21.2 |
| " | 16 | 2.3 | 13.7 |
| 7.5% | 100 | 21.8 | 78.2 |
| " | 75 | 16.2 | 58.8 |
| " | 50 | 11 | 39 |
| " | 25 | 5.5 | 19.5 |
| " | 16 | 3.5 | 12.5 |
| 6% | 100 | 17.4 | 82.6 |
| " | 75 | 13 | 62 |
| " | 50 | 8.7 | 41.3 |
| " | 25 | 4.4 | 20.6 |
| " | 16 | 2.9 | 13.1 |
| 5% | 100 | 14.5 | 85.5 |
| " | 75 | 11 | 64 |
| " | 50 | 7.3 | 42.7 |
| " | 25 | 3.8 | 21.2 |
| " | 16 | 2.3 | 13.7 |
| 2.5% | 100 | 7.3 | 92.7 |
| " | 75 | 5.5 | 69.5 |
| " | 50 | 3.8 | 46.2 |
| " | 25 | 1.7 | 23.3 |
| " | 16 | 1.2 | 14.8 |
| 2% | 100 | 5.8 | 94.2 |

| Strength Desired | Amount Finished Product Desired (Oz.) | Oz. Con. to Add | Oz. Diluent to Add |
|---|---|---|---|
| 2% | 75 | 4.4 | 70.6 |
| " | 50 | 2.9 | 47.1 |
| " | 25 | 1.5 | 23.5 |
| " | 16 | 0.9 | 15.1 |
| 1% | 100 | 2.9 | 97.1 |
| " | 75 | 2.3 | 72.7 |
| " | 50 | 1.5 | 48.5 |
| " | 25 | 0.9 | 24.1 |
| " | 16 | 0.6 | 15.4 |
| 0.5% (approx.) | 100 | 1.5 | 98.5 |
| " (approx.) | 75 | 1.2 | 73.8 |
| " (approx.) | 50 | 0.9 | 49.1 |
| " (approx.) | 25 | 0.4 | 24.6 |
| " (approx.) | 16 | 0.3 | 15.7 |

## 30% Concentrate

| Strength Desired | Amount Finished Product Desired (Oz.) | Oz. Con. to Add | Oz. Diluent to Add |
|---|---|---|---|
| 30% | 100 (3.1 qt.) | 100 | 0 |
| " | 75 (2.3 qt.) | 75 | 0 |
| " | 50 (1.6 qt.) | 50 | 0 |
| " | 25 (1.6 pt.) | 25 | 0 |
| " | 16 (1 pt.) | 16 | 0 |
| 20% | 100 | 66 | 34 |
| " | 75 | 49.5 | 50.5 |
| " | 50 | 33 | 17 |
| " | 25 | 16.5 | 8.5 |
| " | 16 | 10.6 | 5.4 |
| 15% | 100 | 49.5 | 50.5 |
| " | 75 | 37.3 | 37.3 |
| " | 50 | 24.8 | 25.2 |
| " | 25 | 12.5 | 12.5 |
| " | 16 | 7.9 | 8.1 |
| 10% | 100 | 33 | 67 |
| " | 75 | 24.8 | 50.2 |

| Strength Desired | Amount Finished Product Desired (Oz.) | Oz. Con. to Add | Oz. Diluent to Add |
|---|---|---|---|
| " | 50 | 16.5 | 33.5 |
| " | 25 | 8.3 | 16.7 |
| " | 16 | 5.3 | 10.7 |
| 7.5% | 100 | 24.8 | 75.2 |
| " | 75 | 18.5 | 56.5 |
| " | 50 | 12.5 | 37.5 |
| " | 25 | 6.3 | 18.7 |
| " | 16 | 4 | 12 |
| 6% | 100 | 19.8 | 80.2 |
| " | 75 | 14.9 | 60.1 |
| " | 50 | 9.9 | 40.1 |
| " | 25 | 5 | 20 |
| " | 16 | 3.3 | 12.7 |
| 5% | 100 | 16.5 | 83.5 |
| " | 75 | 12.5 | 62.5 |
| " | 50 | 8.3 | 41.7 |
| " | 25 | 4.3 | 20.7 |
| " | 16 | 2.6 | 13.4 |
| 2.5% | 100 | 8.3 | 91.7 |
| " | 75 | 6.3 | 68.7 |
| " | 50 | 4.3 | 45.7 |
| " | 25 | 2 | 23 |
| " | 16 | 1.3 | 14.7 |
| 2% | 100 | 6.6 | 93.4 |
| " | 75 | 5 | 70 |
| " | 50 | 3.3 | 46.7 |
| " | 25 | 1.7 | 23.3 |
| " | 16 | 1 | 15 |
| 1% | 100 | 3.3 | 96.7 |
| " | 75 | 2.6 | 72.4 |
| " | 50 | 1.7 | 48.3 |
| " | 25 | 1 | 24 |
| " | 16 | 0.7 | 15.3 |
| 0.5% | 100 | 1.7 | 98.3 |
| " | 75 | 1.3 | 73.7 |
| " | 50 | 1 | 49 |
| " (approx.) | 25 | 0.3 | 24.7 |
| " (approx.) | 16 | 0.3 | 15.7 |

## 25% Concentrate

| Strength Desired | Amount Finished Product Desired (Oz.) | Oz. Con. to Add | Oz. Diluent to Add |
|---|---|---|---|
| 25% | 100 (3.1 qt.) | 0 | 0 |
| " | 75 (2.3 qt.) | 0 | 0 |
| " | 50 (1.6 qt.) | 0 | 0 |
| " | 25 (1.6 pt.) | 0 | 0 |
| " | 16 (1 pt.) | 0 | 0 |
| 20% | 100 | 80 | 20 |
| " | 75 | 60 | 15 |
| " | 50 | 40 | 10 |
| " | 25 | 20 | 5 |
| " | 16 | 12.8 | 3.2 |
| 15% | 100 | 60 | 40 |
| " | 75 | 45.2 | 29.8 |
| " | 50 | 30 | 20 |
| " | 25 | 15.2 | 9.8 |
| " | 16 | 9.6 | 6.4 |
| 10% | 100 | 40 | 60 |
| " | 75 | 30 | 45 |
| " | 50 | 20 | 30 |
| " | 25 | 10 | 15 |
| " | 16 | 6.4 | 9.6 |
| 7.5% | 100 | 30 | 70 |
| " | 75 | 22.4 | 52.6 |
| " | 50 | 15.2 | 34.8 |
| " | 25 | 7.6 | 17.4 |
| " | 16 | 4.8 | 11.2 |
| 6% | 100 | 24 | 76 |
| " | 75 | 18 | 57 |
| " | 50 | 12 | 38 |
| " | 25 | 6 | 19 |
| " | 16 | 4 | 12 |
| 5% | 100 | 20 | 80 |
| " | 75 | 15.2 | 59.8 |
| " | 50 | 10 | 40 |
| " | 25 | 5.2 | 19.8 |
| " | 16 | 3.2 | 12.8 |
| 2.5% | 100 | 10 | 90 |

| Strength Desired | Amount Finished Product Desired (Oz.) | Oz. Con. to Add | Oz. Diluent to Add |
|---|---|---|---|
| " | 75 | 7.6 | 67.4 |
| " | 50 | 5.2 | 44.8 |
| " | 25 | 2.4 | 22.6 |
| " | 16 | 1.6 | 14.4 |
| 2% | 100 | 8 | 92 |
| " | 75 | 6 | 69 |
| " | 50 | 4 | 46 |
| " | 25 | 2 | 23 |
| " | 16 | 1.2 | 13.8 |
| 1% | 100 | 4 | 96 |
| " | 75 | 3.2 | 71.8 |
| " | 50 | 2 | 48 |
| " | 25 | 1.2 | 23.8 |
| " | 16 | 0.8 | 15.2 |
| 0.5% | 100 | 2 | 98 |
| " | 75 | 1.6 | 73.4 |
| " | 50 | 1.2 | 48.8 |
| " (approx.) | 25 | 0.5 | 24.5 |
| " (approx.) | 16 | 0.4 | 15.6 |

## 20% Concentrate

| Strength Desired | Amount Finished Product Desired (Oz.) | Oz. Con. to Add | Oz. Diluent to Add |
|---|---|---|---|
| 20% | 100 (3.1 qt.) | 0 | 0 |
| " | 75 (2.3 qt.) | 0 | 0 |
| " | 50 (1.6 qt.) | 0 | 0 |
| " | 25 (1.6 pt.) | 0 | 0 |
| " | 16 ( 1 pt.) | 0 | 0 |
| 15% | 100 | 75 | 25 |
| " | 75 | 56.5 | 18.5 |
| " | 50 | 37.5 | 12.5 |
| " | 25 | 19 | 6 |
| " | 16 | 12 | 4 |
| 10% | 100 | 50 | 50 |
| " | 75 | 37.5 | 37.5 |

| Strength Desired | Amount Finished Product Desired (Oz.) | Oz. Con. to Add | Oz. Diluent to Add |
|---|---|---|---|
| 10% | 50 | 25 | 25 |
| " | 25 | 12.5 | 12.5 |
| " | 16 | 8 | 8 |
| 7.5% | 100 | 37.5 | 62.5 |
| " | 75 | 28 | 47 |
| " | 50 | 19 | 31 |
| " | 25 | 9.5 | 15.5 |
| " | 16 | 6 | 10 |
| 6% | 100 | 30 | 70 |
| " | 75 | 22.5 | 52.5 |
| " | 50 | 15 | 35 |
| " | 25 | 7.5 | 17.5 |
| " | 16 | 5 | 11 |
| 5% | 100 | 25 | 75 |
| " | 75 | 19 | 56 |
| " | 50 | 12.5 | 47.5 |
| " | 25 | 6.5 | 18.5 |
| " | 16 | 4 | 12 |
| 2.5% | 100 | 12.5 | 87.5 |
| " | 75 | 9.5 | 65.5 |
| " | 50 | 6.5 | 43.5 |
| " | 25 | 3 | 22 |
| " | 16 | 2 | 14 |
| 2% | 100 | 10 | 90 |
| " | 75 | 7.5 | 67.5 |
| " | 50 | 5 | 45 |
| " | 25 | 2.5 | 22.5 |
| " | 16 | 1.5 | 13.5 |
| 1% | 100 | 5 | 95 |
| " | 75 | 4 | 71 |
| " | 50 | 2.5 | 47.5 |
| " | 25 | 1.5 | 23.5 |
| " | 16 | 1 | 15 |
| 0.5% | 100 | 2.5 | 97.5 |
| " | 75 | 2 | 73 |
| " (approx.) | 50 | 1.5 | 48.5 |
| " (approx.) | 25 | 0.6 | 24.4 |
| " | 16 | 0.5 | 15.5 |

## DUST OR POWDER PESTICIDES

In most cases, dust and powder pesticides also will be concentrated and, thus, should be diluted before use. In general, mixing dusts or powders with their diluents or carriers, which also are dry materials, will prove more difficult than mixing liquids. Dusts and powder pesticides must be mixed in a weight-to-weight ratio. That is, you must *weigh out* the amount of dust concentrate to be mixed with a given weight of carrier or diluent. To do this, you will need a balance, or scales, capable of weighing in *ounces*.

What do you mix with concentrated dust or powder pesticides? The list below includes the most-commonly used and recommended diluents or carriers for pesticide dusts and powders. It is suggested that you use the one (or ones) that are most easily available or preferable to you.

### Diluent/Carriers for Pesticide Dusts and Powders

| | |
|---|---|
| Clay | Gypsum |
| Fuller's earth | Pyrophyllite |
| Frianite | Talc |

### 100% Dust or Powder Concentrate

| Strength Desired | Oz. Finished Product Desired | Oz. Con. to Add | Oz. Diluent to Add |
|---|---|---|---|
| 20% | 48 (3 lb.) | 9.6 | 38.4 |
| " | 32 (2 lb.) | 6.4 | 25.6 |
| " | 16 (1 lb.) | 3.2 | 12.8 |
| " | 8 (½ lb.) | 1.6 | 6.4 |
| 15% | 48 | 7.2 | 40.8 |
| " | 32 | 4.8 | 27.2 |
| " | 16 | 2.4 | 13.6 |
| " | 8 | 1.2 | 6.8 |
| 12% | 48 | 5.7 | 42.3 |
| " | 32 | 3.8 | 28.2 |
| " | 16 | 1.9 | 14.1 |
| " | 8 | (approx.) 1 | 7 |
| 10% | 48 | 4.8 | 43 |
| " | 32 | 3.2 | 28.8 |
| " | 16 | 1.6 | 14.4 |
| " | 8 | 0.8 | 7.2 |

| Strength Desired | Oz. Finished Product Desired | Oz. Con. to Add | Oz. Diluent to Add |
|---|---|---|---|
| 7% | 48 | 3.4 | 44.6 |
| " | 32 | 2.2 | 29.8 |
| " | 16 | 1.1 | 14.9 |
| " | 8 | 0.6 | 7.4 |
| 5% | 48 | 2.4 | 45.6 |
| " | 32 | 1.6 | 30.4 |
| " | 16 | 0.8 | 15.2 |
| " | 8 | 0.4 | 7.6 |
| 3% | 48 | 1.4 | 46.6 |
| " | 32 | 1 | 33 |
| " | 16 | 0.5 | 15.5 |
| " | 8 | 0.2 | 7.8 |
| 2% | 48 | 1 | 47 |
| " | 32 | 0.6 | 31.4 |
| " | 16 | 0.3 | 15.7 |
| " | 8 | 0.08 | (approx.) 7.9 |
| 1% | 48 | 0.5 | (approx.) 47.5 |
| " | 32 | 0.3 | (approx.) 31.7 |
| " | 16 | 0.17 | (approx.) 15.8 |
| " | 8 | 0.08 | (approx.) 7.9 |
| 0.5% | 48 | (approx.) 0.2 | (approx.) 47.8 |
| " | 32 | (approx.) 0.16 | (approx.) 31.8 |
| " | 16 | (approx.) 0.08 | (approx.) 15.9 |
| " | 8 | (approx.) 0.04 | (approx.) 7.9 |

## 75% Dust or Powder Concentrate

| Strength Desired | Oz. Finished Product Desired | Oz. Con. to Add | Oz. Diluent to Add |
|---|---|---|---|
| 20% | 48 (3 lb.) | 12 | 36 |
| " | 32 (2 lb.) | 8 | 24 |
| " | 16 (1 lb.) | 4 | 12 |
| " | 8 (½ lb.) | 2 | 6 |
| 15% | 48 | 9 | 39 |
| " | 32 | 6 | 26 |
| " | 16 | 3 | 13 |
| " | 8 | 1.5 | 6.5 |
| 12% | 48 | 7.1 | 40.9 |

| Strength Desired | Oz. Finished Product Desired | Oz. Con. to Add | Oz. Diluent to Add |
|---|---|---|---|
| 12% | 32 | 4.8 | 27.2 |
| " | 16 | 2.4 | 13.6 |
| " | 8 | (approx.) 1.3 | (approx.) 6.7 |
| 10% | 48 | 6 | 42 |
| " | 32 | 4 | 28 |
| " | 16 | 2 | 14 |
| " | 8 | 1 | 7 |
| 7% | 48 | 4.3 | 43.7 |
| " | 32 | 2.8 | 29.2 |
| " | 16 | 1.4 | 14.6 |
| " | 8 | (approx.) 0.7 | (approx.) 7.3 |
| 5% | 48 | 3 | 45 |
| " | 32 | 2 | 30 |
| " | 16 | 1 | 15 |
| " | 8 | (approx.) 0.5 | 7.5 |
| 3% | 48 | 1.8 | 46.2 |
| " | 32 | 1.3 | 30.7 |
| " | 16 | (approx.) 0.6 | 15.4 |
| " | 8 | (approx.) 0.3 | 7.7 |
| 2% | 48 | 1.3 | 46.7 |
| " | 32 | 0.8 | 31.2 |
| " | 16 | (approx.) 0.4 | 15.6 |
| " | 8 | (approx.) 0.2 | 7.8 |
| 1% | 48 | 0.6 | 47.4 |
| " | 32 | 0.4 | 31.6 |
| " | 16 | (approx.) 0.2 | 15.8 |
| " | 8 | (approx.) 0.1 | 7 |
| 0.5% | 48 | (approx.) 0.25 | 48 |
| " | 32 | (approx.) 0.2 | 39 |
| " | 16 | (approx.) 0.1 | 15 |
| " | 8 | (approx.) 0.05 | 8 |

## 50% Dust or Powder Concentrate

| Strength Desired | Oz. Finished Product Desired | Oz. Con. to Add | Oz. Diluent to Add |
|---|---|---|---|
| 20% | 48 (3 lb.) | 19.2 | 28.8 |
| " | 32 (2 lb.) | 12.8 | 19.2 |

| Strength Desired | Oz. Finished Product Desired | Oz. Con. to Add | Oz. Diluent to Add |
|---|---|---|---|
| 20% | 16 (1 lb.) | 6.4 | 9.6 |
| " | 8 (½ lb.) | 3.2 | 4.8 |
| 15% | 48 | 14.4 | 33.6 |
| " | 32 | 9.6 | 22 |
| " | 16 | 4.8 | 11 |
| " | 8 | 2.4 | 5.6 |
| 12% | 48 | 11 | 37 |
| " | 32 | 7.6 | 24 |
| " | 16 | 3.8 | 12 |
| " | 8 | 2 | 6 |
| 10% | 48 | 9.6 | 38 |
| " | 32 | 6.4 | 26 |
| " | 16 | 3.2 | 13 |
| " | 8 | 1.6 | 6 |
| 7% | 48 | 6.8 | 41 |
| " | 32 | 4.4 | 32 |
| " | 16 | 2.2 | 16 |
| " | 8 | 1.1 | 7 |
| 5% | 48 | 4.8 | 43 |
| " | 32 | 3.2 | 29 |
| " | 16 | 1.6 | 14 |
| " | 8 | 0.8 | 7 |
| 3% | 48 | 2.8 | 45 |
| " | 32 | 2 | 30 |
| " | 16 | 1 | 15 |
| " | 8 | 0.4 | 8 |
| 2% | 48 | 2 | 46 |
| " | 32 | 1.2 | 31 |
| " | 16 | 0.6 | 15 |
| " | 8 | 0.3 | 8 |
| 1% | 48 | 1 | 47 |
| " | 32 | 0.6 | 31 |
| " | 16 | 0.3 | 16 |
| " | 8 | 0.2 | 8 |
| 0.5% | 48 | 0.4 | 48 |
| " | 32 | 0.32 | 32 |
| " | 16 | 0.2 | 16 |
| " | 8 | 0.1 | 8 |

## 35% **Dust or Powder Concentrate**

| Strength Desired | Oz. Finished Product Desired | Oz. Con. to Add | Oz. Diluent to Add |
|---|---|---|---|
| 20% | 48 (3 lb.) | 27 | 21 |
| " | 32 (2 lb.) | 18 | 14 |
| " | 16 (1 lb.) | 9 | 7 |
| " | 8 (½ lb.) | 5 | 3 |
| 15% | 48 | 21 | 28 |
| " | 32 | 14 | 18 |
| " | 16 | 7 | 9 |
| " | 8 | 3 | 5 |
| 12% | 48 | 16 | 32 |
| " | 32 | 11 | 21 |
| " | 16 | 5 | 11 |
| " | 8 | 3 | 5 |
| 10% | 48 | 14 | 34 |
| " | 32 | 9 | 23 |
| " | 16 | 5 | 11 |
| " | 8 | 2 | 6 |
| 7% | 48 | 10 | 38 |
| " | 32 | 6 | 26 |
| " | 16 | 3 | 13 |
| " | 8 | 2 | 6 |
| 5% | 48 | 7 | 41 |
| " | 32 | 5 | 27 |
| " | 16 | 2 | 14 |
| " | 8 | 1 | 7 |
| 3% | 48 | 4 | 44 |
| " | 32 | 3 | 29 |
| " | 16 | 1 | 15 |
| " | 8 | 0.6 | 7 |
| 2% | 48 | 3 | 45 |
| " | 32 | 2 | 30 |
| " | 16 | 1 | 15 |
| " | 8 | 0.2 | 8 |
| 1% | 48 | 1.4 | 47 |
| " | 32 | 1 | 31 |
| " | 16 | 0.5 | 16 |
| " | 8 | 0.2 | 8 |

| Strength Desired | Oz. Finished Product Desired | Oz. Con. to Add | Oz. Diluent to Add |
|---|---|---|---|
| 0.5% | 48 | 0.6 | 47 |
| " | 32 | 0.5 | 31 |
| " | 16 | 0.2 | 16 |
| " | 8 | 0.1 | 8 |

## 30% Dust or Powder Concentrate

| Strength Desired | Oz. Finished Product Desired | Oz. Con. to Add | Oz. Diluent to Add |
|---|---|---|---|
| 20% | 48 (3 lb.) | 32 | 16 |
| " | 32 (2 lb.) | 21 | 11 |
| " | 16 (1 lb.) | 11 | 5 |
| " | 8 (½ lb.) | 5 | 3 |
| 15% | 48 | 24 | 24 |
| " | 32 | 16 | 16 |
| " | 16 | 8 | 8 |
| " | 8 | 4 | 4 |
| 12% | 48 | 19 | 29 |
| " | 32 | 13 | 19 |
| " | 16 | 6 | 10 |
| " | 8 | 3 | 5 |
| 10% | 48 | 16 | 32 |
| " | 32 | 11 | 21 |
| " | 16 | 5 | 11 |
| " | 8 | 3 | 5 |
| 7% | 48 | 11 | 37 |
| " | 32 | 7 | 25 |
| " | 16 | 4 | 12 |
| " | 8 | 2 | 6 |
| 5% | 48 | 8 | 40 |
| " | 32 | 5 | 27 |
| " | 16 | 3 | 13 |
| " | 8 | 1 | 7 |
| 3% | 48 | 5 | 43 |
| " | 32 | 3 | 29 |
| " | 16 | 2 | 14 |
| " | 8 | 1 | 7 |

| Strength Desired | Oz. Finished Product Desired | Oz. Con. to Add | Oz. Diluent to Add |
|---|---|---|---|
| 2% | 48 | 3 | 45 |
| " | 32 | 2 | 30 |
| " | 16 | 1 | 15 |
| " | 8 | 0.5 | 8 |
| 1% | 48 | 2 | 46 |
| " | 32 | 1 | 31 |
| " | 16 | 0.5 | 16 |
| " | 8 | 0.3 | 8 |
| 0.5% | 48 | 1 | 47 |
| " | 32 | 0.5 | 32 |
| " | 16 | 0.3 | 16 |
| " | 8 | 0.2 | 8 |

## 25% Dust or Powder Concentrate

| Strength Desired | Oz. Finished Product Desired | Oz. Con. to Add | Oz. Diluent to Add |
|---|---|---|---|
| 20% | 48 (3 lb.) | 36 | 12 |
| " | 32 (2 lb.) | 27 | 5 |
| " | 16 (1 lb.) | 13 | 3 |
| " | 8 (½ lb.) | 6 | 2 |
| 15% | 48 | 29 | 19 |
| " | 32 | 19 | 13 |
| " | 16 | 10 | 6 |
| " | 8 | 5 | 3 |
| 12% | 48 | 23 | 25 |
| " | 32 | 15 | 17 |
| " | 16 | 8 | 8 |
| " | 8 | 4 | 4 |
| 10% | 48 | 19 | 29 |
| " | 32 | 13 | 19 |
| " | 16 | 6 | 10 |
| " | 8 | 3 | 5 |
| 7% | 48 | 14 | 34 |
| " | 32 | 9 | 23 |
| " | 16 | 4 | 12 |
| " | 8 | 2 | 6 |

| Strength Desired | Oz. Finished Product Desired | Oz. Con. to Add | Oz. Diluent to Add |
|---|---|---|---|
| 5% | 48 | 10 | 38 |
| " | 32 | 6 | 24 |
| " | 16 | 3 | 13 |
| " | 8 | 2 | 6 |
| 3% | 48 | 6 | 42 |
| " | 32 | 4 | 28 |
| " | 16 | 2 | 14 |
| " | 8 | 1 | 7 |
| 2% | 48 | 4 | 44 |
| " | 32 | 2 | 30 |
| " | 16 | 1 | 15 |
| " | 8 | 0.5 | 8 |
| 1% | 48 | 2 | 46 |
| " | 32 | 1 | 31 |
| " | 16 | 0.5 | 16 |
| " | 8 | 0.3 | 8 |
| 0.5% | 48 | 1 | 47 |
| " | 32 | 0.6 | 31 |
| " | 16 | 0.3 | 16 |
| " | 8 | 0.2 | 8 |

## 20% Dust or Powder Concentrate

| Strength Desired | Oz. Finished Product Desired | Oz. Con. to Add | Oz. Diluent to Add |
|---|---|---|---|
| 20% | 48 (3 lb.) | 48 | 0 |
| " | 32 (2 lb.) | 32 | 0 |
| " | 16 (1 lb.) | 16 | 0 |
| " | 8 (½ lb.) | 8 | 0 |
| 15% | 48 | 36 | 12 |
| " | 32 | 24 | 8 |
| " | 16 | 12 | 4 |
| " | 8 | 6 | 2 |
| 12% | 48 | 29 | 19 |
| " | 32 | 19 | 13 |
| " | 16 | 10 | 6 |
| " | 8 | 5 | 3 |

| Strength Desired | Oz. Finished Product Desired | Oz. Con. to Add | Oz. Diluent to Add |
|---|---|---|---|
| 10% | 48 | 24 | 24 |
| " | 32 | 16 | 16 |
| " | 16 | 8 | 8 |
| " | 8 | 4 | 4 |
| 7% | 48 | 17 | 31 |
| " | 32 | 11 | 21 |
| " | 16 | 6 | 10 |
| " | 8 | 3 | 5 |
| 5% | 48 | 12 | 36 |
| " | 32 | 8 | 24 |
| " | 16 | 4 | 12 |
| " | 8 | 2 | 6 |
| 3% | 48 | 7 | 41 |
| " | 32 | 5 | 27 |
| " | 16 | 3 | 13 |
| " | 8 | 2 | 7 |
| 2% | 48 | 5 | 43 |
| " | 32 | 3 | 29 |
| " | 16 | 2 | 15 |
| " | 8 | 1 | 7 |
| 1% | 48 | 3 | 46 |
| " | 32 | 2 | 31 |
| " | 16 | 1 | 15 |
| " | 8 | 0.5 | 8 |
| 0.5% | 48 | 1 | 47 |
| " | 32 | 0.8 | 31 |
| " | 16 | 0.4 | 16 |
| " | 8 | 0.2 | 8 |

# BIBLIOGRAPHY

*Annual pesticide recommendations.* 1980–81. Residex Corp., Clark, New Jersey.

Arnett, Jr., Ross H. 1968. *The beetles of the United States.* The American Entomological Institute, Ann Arbor.

Borror, Donald J., and White, Richard E. 1970. *A field guide to the insects of America north of Mexico.* Boston: Houghton Mifflin Co.

Borror, Donald J., DeLong, Dwight M., and Triplehorn, Charles A. 1976. *An introduction to the study of insects.* New York: Holt, Rinehart & Winston.

Brooks, Joe E., and Peck, Thomas D., eds. 1969. *Community pest and related vector control.* Pest Control Operators of California, Inc., Los Angeles.

Comstock, John H. 1949. *An introduction to entomology.* Ithaca: Comstock Publishing Co.

Ebeling, Walter. 1975. *Urban entomology.* University of California, Div. of Agricultural Sciences.

Extension Bulletin no. 387. Rev. 1980. Agricultural Extension Service, University of Minnesota. St. Paul, Minnesota.

Extension Bulletin no. 412. Rev. 1980. Agricultural Extension Service, University of Minnesota. St. Paul, Minnesota.

Extension Folder no. 414. Rev. 1980. Agricultural Extension Service, University of Minnesota. St. Paul, Minnesota.

Horn, David J. 1978. *Biology of insects.* Philadelphia: W. B. Saunders Co.

James, Maurice T., and Harwood, Robert F. 1969. *Herms's medical entomology.* 6th ed. London: Collier-Macmillan Ltd.

McKillip, Barbara B., ed. 1973. *Getting the Bugs out of Organic Gardening and Farming.* Emmaus, Pennsylvania: Rodale Press.

Mallis, Arnold. 1969. *Handbook of pest control.* 5th ed. New York: Macnair-Dorland Co.

*Mosquitoes of public health importance and their control.* 1960. U.S. Department of Health, Education, and Welfare, Public Health Service, Atlanta.

Pratt, Harry D., and Brown, Robert Z. 1977. *Biological factors in domestic rodent control.* U.S. Department of Health, Education, and Welfare, Public Health Service, Atlanta.

Pratt, Harry D., and Littig, Kent S. 1974. *Insecticide application equipment for the control of insects of public health importance.* U.S. Department of Health, Education, and Welfare, Public Health Service, Atlanta.

Pratt, Harry D., and Littig, Kent S. 1974. *Insecticides for the control of insects of public health importance.* U.S. Department of Health, Education, and Welfare, Public Health Service, Atlanta.

Pratt, Harry D., and Littig, Kent S. 1960. *Introduction to arthropods of public health importance.* U.S. Department of Health, Education, and Welfare, Public Health Service, Atlanta.

Pratt, Harry D., and Littig, Kent S. 1973. *Lice of public health importance and their control.* U.S. Department of Health, Education, and Welfare, Public Health Service, Atlanta.

Pratt, Harry D., and Littig, Kent S. 1962. *Ticks of public health importance and their control.* U.S. Department of Health, Education, and Welfare, Public Health Service, Atlanta.

Pratt, Harry D., and Stark, Harold E. 1973. *Fleas of public health importance and their control.* U.S. Department of Health, Education, and Welfare, Public Health Service, Atlanta.

Pratt, Harry D., Bjornson, Bayard F., and Littig, Kent S. 1977. *Control of domestic rats and mice.* U.S. Department of Health, Education, and Welfare, Public Health Service, Atlanta.

Pratt, Harry D., Littig, Kent S., and Scott, Harold G. 1975. *Household and stored-food insects of public health importance and their control.* U.S. Department of Health, Education, and Welfare, Public Health Service, Atlanta.

Pratt, Harry D., Littig, Kent S., and Scott, Harold G. 1975. *Flies of public health importance and their control.* U.S. Department of Health, Education, and Welfare, Public Health Service, Atlanta.

Quraishi, M. Sayeed. 1977. *Biochemical insect control.* New York: John Wiley and Sons.

Rice, Paul L., and Pratt, Harry D. 1978. *Epidemiology and control of vectorborne diseases.* U.S. Department of Health, Education, and Welfare, Public Health Service, Atlanta.

Scott, Harold S., and Pratt, Harry D. 1959. *Scorpions, spiders, and other arthropods of minor public health importance and their control.* U.S. Department of Health, Education, and Welfare, Public Health Service, Atlanta.

Sharp, David. *Insects.* Vols. I and II. 1970. New York: Dover Publications, Inc.

Truman, Lee C., Bennett, Gary W., and Butts, William L. 1976. *Scientific guide to pest control operations.* 3rd ed. Cleveland: Harvest Publishing Co.

Ware, George W. 1975. *Pesticides, an auto-tutorial approach.* San Franscisco: W. H. Freeman and Co.

Westcott, Cynthia. 1946. *The Gardener's Bug Book.* New York: Doubleday and Co., Inc.